Editorial Manager	Chester Fisher
Senior Editor	Lynne Sabel
Editor	John Rowlstone
Assistant Editor	Bridget Daly
Series Designers	QED (Alaistair Campbell and Edward Kinsey)
Designer	Mike Blore
Series Consultant	Keith Lye
Consultant	John Stidworthy
Production	Penny Kitchenham
Picture Research	Jenny Golden

© Macdonald Educational Ltd. 1978
First published 1978
Reprinted 1979
Macdonald Educational
Holywell House
Worship Street
London EC2A 2EN

2081/3200
ISBN 0 356 05759 3

Designed and created in
Great Britain

Printed and bound by
New Interlitho, Italy

The Living World

Cathy Kilpatrick and
Mark Lambert

Macdonald

Contents

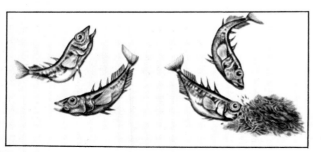

The Plant World

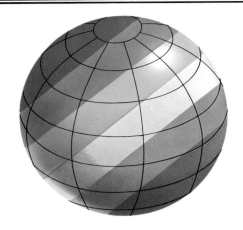

World of Knowledge

This book breaks new ground in the method it uses to present information to the reader. The unique page design combines narrative with an alphabetical reference section and it uses colourful photographs, diagrams and illustrations to provide an instant and detailed understanding of the book's theme. The main body of information is presented in a series of chapters that cover, in depth, the subject of this book. At the bottom of each page is a reference section which gives, in alphabetical order, concise articles which define, or enlarge on, the topics discussed in the chapter. Throughout the book, the use of SMALL CAPITALS in the text directs the reader to further information that is printed in the reference section. The same method is used to cross-reference entries within each reference section. Finally, there is a comprehensive index at the end of the book that will help the reader find information in the text, illustrations and reference sections. The quality of the text, and the originality of its presentation, ensure that this book can be read both for enjoyment and for the most up-to-date information on the subject.

WORLD OF KNOWLEDGE

The Animal World

Cathy Kilpatrick

Introduction

The Animal World is a broad survey of the animal kingdom and animal behaviour in all its fascinating variety, both on land, in the sea and in the air. There are more than 1,000,000 species of animals, not including the enormous number of extinct species which we know about only from their fossil remains. Living species range from the blue whale, which is probably the largest and heaviest animal that has ever lived on Earth, to minute protozoans which can be seen only through a microscope. **The Animal World** also shows how each species has a special relationship with other animals and plants around it and how each species has evolved particular adaptations to its environment which enable it to live and breed successfully. However, should this environment be destroyed or altered, especially by the activities of man, entire species can be threatened with extinction.

The land, sea and air are filled with a tremendous variety of living animals. All these animals, which differ greatly in size and shape, are specially adapted so that they can survive and breed successfully.

Animals Great and Small

There are more than one million different kinds, or species, of living animals on the planet Earth, ranging in almost infinite variety from domestic dogs and cats to wild tigers and elephants and birds, fish, beetles and butterflies. New species are being discovered all the time by scientists and naturalists. This enormous variety of animal life has evolved many different ways of moving about from place to place, such as walking, crawling, swimming, hopping, flying, gliding and burrowing.

An animal usually adapts to living in a particular habitat in its ENVIRONMENT. Living creatures can usually be found in almost every part of the Earth. For example, polar bears combat the same inhospitable climate of the north polar regions as do penguins at the South Pole. In the marine environment, certain fishes are equipped for life at the bottom of the ocean and are able to withstand the terrific pressure of the waters there. Some animals are adapted to live on or inside the bodies of animals and are known as PARASITES, while others actually form a mutually beneficial association known as symbiosis.

Man is the most intelligent species in the ANIMAL KINGDOM. He has certainly adapted himself to live in almost all the environments found on Earth and dominates life today. However, when we study the various numbers of species in the main groups, we discover, amazingly, that the greatest variety of species is not in man's group, the mammals, but in the Arthropoda, a group of the INVERTEBRATES. The Arthropoda includes beetles, crabs, butterflies, wasps and flies. Over 900,000 species are identified in this group, all of them having hard, horny skeletons on the outside of their bodies. Fishes, birds, amphibians, reptiles and mammals, all of which are VERTEBRATES (also known as CHORDATES), account for only 45,000 species. Perhaps more

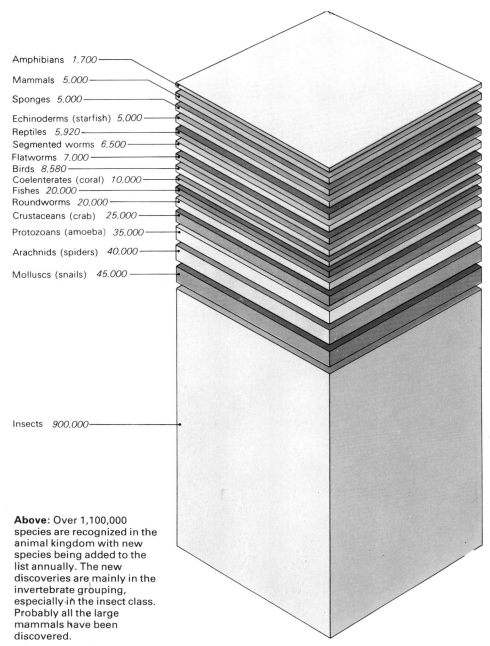

Amphibians *1,700*
Mammals *5,000*
Sponges *5,000*
Echinoderms (starfish) *5,000*
Reptiles *5,920*
Segmented worms *6,500*
Flatworms *7,000*
Birds *8,580*
Coelenterates (coral) *10,000*
Fishes *20,000*
Roundworms *20,000*
Crustaceans (crab) *25,000*
Protozoans (amoeba) *35,000*
Arachnids (spiders) *40,000*
Molluscs (snails) *45,000*
Insects *900,000*

Above: Over 1,100,000 species are recognized in the animal kingdom with new species being added to the list annually. The new discoveries are mainly in the invertebrate grouping, especially in the insect class. Probably all the large mammals have been discovered.

Reference

A **Animal kingdom** includes all the known species of animals in the world. The kingdom is split up into various divisions in a system of CLASSIFICATION.
Aristotle (384-322 BC) was one of the early Greek thinkers. In his philosophical works on biology he showed the importance of change in nature. He understood that the speck of matter in a hen's egg becomes, by definite stages, a chicken.

C **Cambrian period** (570-500 million years ago). Many FOSSILS are found in these earliest rocks.
Carboniferous period

Wallaby with young in pouch

(345-280 million years ago). The climate in many parts of the world was then warm and moist. Dense, swampy forests were found on land.

Chordata are all the animals that possess a notochord (the forerunner to the backbone). Among the first chordates were the graptolites, an extinct group of marine animals which thrived in the CAMBRIAN PERIOD.
Classification is the system in biology by which organisms are arranged into a series of groups within groups according to certain anatomical characteristics. The smallest unit is a species. This includes individuals that are biologically the same and will interbreed

naturally. Groups of similar species form a genus, with groups of genera forming a family. Families form an order, orders form a class, classes form a phylum, and phyla form a kingdom. The system was devised by Carolus Linnaeus (1707-78), a Swedish biologist.
Cretaceous period (135-65 million years ago). Its name comes from the Latin *creta*, meaning chalk. Many of the DINOSAURS evolved during the Cretaceous period
Cuvier, Georges (1769-1832), rejected all ideas of changes in animal form and

surprising is that of over one million species of animals, some 95 per cent have no internal skeleton at all, many of them also being invisible to the human eye.

Evolution of life

The modern diversity of animal life is a result of over 500 million years of evolutionary development. For generations, NATURAL SELECTION has favoured the individuals best adapted to their way of life by eliminating their inferior competitors. This theory was first put forward by Charles DARWIN and Alfred WALLACE in the 19th century. Evolution is still at work today. Animal species are continually adapting to their changing surroundings or environment. Peppered moths originally evolved to live on lichen-covered trees as they offered a protective camouflage. The moth's wings have since darkened in colour to match a new environment – that of the polluted towns and countryside.

Early theories about the origin of life now seem rather amusing. Thales (640–546 BC) believed water was the origin of life, while Anaximenes (c. 500 BC) said all things came from the air. Two other strange theories were once believed; that orchids gave birth to birds and small men, and that birds were born from flying fish, lions from sea lions and men from mermen. Early students of theology accepted the theory of creation as told in the book of *Genesis* in the Bible. It was not until the late 17th and 18th centuries that scientists doubted these early theories and advanced new ones. Martin Lister (1638–1712) and Robert Hooke (1635–1703) discovered a great deal by linking the fossils they found with different rock layers. The first startling new theory put forward was probably in 1788, when James Hutton (1726–97) published *Theory of the Earth* explaining the past history of the earth through 'UNIFORMITARIANISM'. Carolus Linnaeus (1707–78), the Swedish biologist, grouped the species into a convenient CLASSIFICATION system, while Jean LAMARCK (1744–1829) was the first person to tackle how species originated.

Today we have an excellent idea of how evolution has taken place with many kinds of evidence helping to build up a picture of the past. It seems to have begun in the sea and progressed onto the land in a series of phases when certain animal groups, for example the amphibians,

became dominant for a time. The amphibians were later eclipsed by the more adaptable reptiles. The history of present-day species has been built up through PALAEONTOLOGY (the study of fossils). Some living forms, such as the king crab and scorpion, are almost identical to their relatives that lived millions of years ago. Perhaps the most celebrated of these LIVING FOSSILS is the coelacanth. This lobe-finned fish was thought to be completely extinct until a specimen was caught in 1938.

Over half the animals that have ever lived are already EXTINCT. Fossil evidence is all that remains of them. The DINOSAURS, PTEROSAURS and mammoths (hairy elephants with long tusks) proved unable to adapt quickly enough to the changing conditions and so died out.

The Australian marsupials have undergone their own private evolution. In most continents the placental mammals proved to be too much competition for the marsupials, but in the relative isolation of Australia they have evolved to fill the ecological roles taken up elsewhere by non-marsupial species. This process is known as adaptive radiation.

Above: The Tuatara lives in New Zealand. It is something of a living fossil, for creatures very similar to it existed about 200 million years ago in the Triassic period.

Above: Ammonites are a wholly extinct class of molluscs known from the early Devonian to late Cretaceous times. A diverse group, some species measured over a metre in diameter.

---

stated that life had been repeatedly wiped out and replaced by new species. This theory was known as catastrophism.

D **Darwin,** Charles (1809-82), was the man who put forward in the 1850s the theory of NATURAL SELECTION and the survival of the fittest to explain how animals have come to change and evolve through the ages. He accompanied the HMS *Beagle* on a voyage to South America and Australasia in 1831 as an unpaid naturalist. When Darwin read the essay *The*

1870s cartoon of Darwin

principles of population by Thomas Malthus, he realized that under the very competitive conditions in the animal and plant world, any variations of species which continued would have to be those which increased the organism's ability to leave fertile offspring. The variations which decreased the animals' or plants' numbers would eventually be eliminated and thus 'natural selection' took place. Darwin revealed his thoughts, and those of WALLACE, in a paper to the Linnaean Society in July 1858. In the following

year he published *The Origin of Species.*
Devonian period (395-345 million years ago). Fishes were dominant in the seas, and it was from the species with lungs and limb-like paired fins that amphibians evolved late in this period.
Dinosaurs were the largest land animals that have lived on the Earth. No man has seen any of these prehistoric monsters alive as they died out 65 million years ago. FOSSILS tell us what they looked like. There were 2 main orders, *Saurischia* and *Ornithischia.* Many di-

nosaurs walked on their hind legs and had massive tails in order to maintain their balance when moving. The crocodile family is the surviving group most closely related to dinosaurs.

E **Embryology** covers an animal's progress from the fertilization of an ovum (or egg) up until birth or hatching.
Environment is the term used to describe the conditions in which an organism lives. These include temperature, light and water, as well as other organisms.

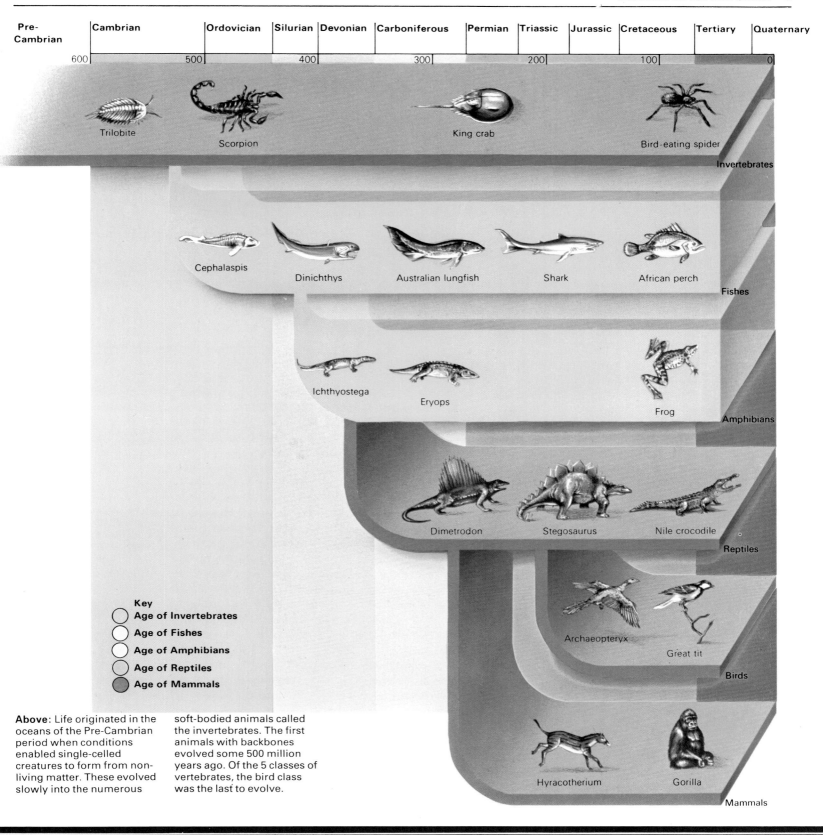

Pre-Cambrian	Cambrian	Ordovician	Silurian	Devonian	Carboniferous	Permian	Triassic	Jurassic	Cretaceous	Tertiary	Quaternary

600   500   400   300   200   100   0

Trilobite   Scorpion   King crab   Bird-eating spider

Invertebrates

Cephalaspis   Dinichthys   Australian lungfish   Shark   African perch

Fishes

Ichthyostega   Eryops   Frog

Amphibians

Dimetrodon   Stegosaurus   Nile crocodile

Reptiles

Key

Age of Invertebrates
Age of Fishes
Age of Amphibians
Age of Reptiles
Age of Mammals

Archaeopteryx   Great tit

Birds

Hyracotherium   Gorilla

Mammals

Above: Life originated in the oceans of the Pre-Cambrian period when conditions enabled single-celled creatures to form from non-living matter. These evolved slowly into the numerous soft-bodied animals called the invertebrates. The first animals with backbones evolved some 500 million years ago. Of the 5 classes of vertebrates, the bird class was the last to evolve.

Evolution means 'unfolding' and describes the gradual process by which all the life forms today have evolved from earlier forms which lived on the Earth millions of years ago. Our present views are based on the theories of Charles DARWIN and Gregor MENDEL.

Tyrannosaurus Rex

Extinction is the term used when an animal or plant dies out. This is the result when the population of a certain species loses more members than it gains through reproduction. This might be brought about by natural changes in the animal's environment, such as climatic changes, that it cannot adapt to. More recently the changes have been caused by man's activities and interference.

F **Fossils** are the hardened remains of plants or animals, or impressions of their

Great Auk (extinct)

forms, preserved in rocks. Generally, only the hard parts like skeletons are preserved. They may be partly or wholly replaced by minerals deposited from circulating water. Fossils are formed in many ways. A common method is when a dead organism is covered by mud or sand, which later changes into rocks; this process can take millions of years.

H **Habitat** is a place or particular ENVIRONMENT, such as a sea shore or a woodland, in which an animal or plant lives.

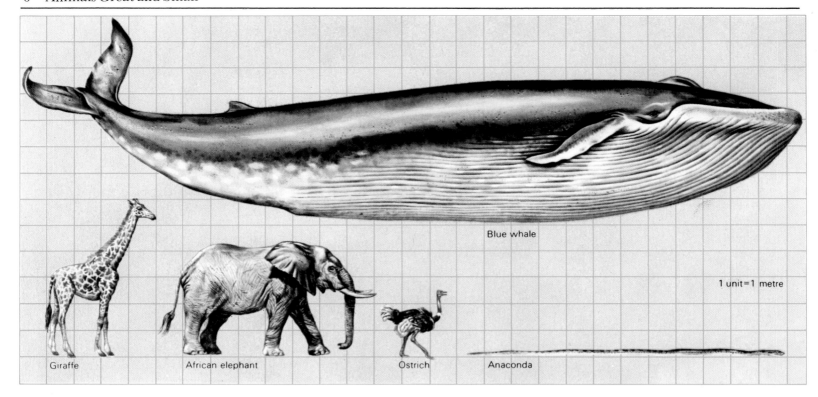

Blue whale

1 unit = 1 metre

Giraffe African elephant Ostrich Anaconda

The record holders

The enormous variety in the animal kingdom is best shown in the records of the largest and smallest animals in a group. Most whales are quite large but the largest and heaviest marine mammal is the blue whale. This species is also known as the sulphur-bottom whale because of the yellow film of microscopic plants (diatoms) often found on its underside. A blue whale is at its heaviest after the summer when it has fed itself on krill (tiny shrimp-like animals) and put on lots of blubber for the winter months when food is scarce. An African elephant put next to a blue whale looks quite dwarfed. In fact, about ten African elephants could stand head to tail on a blue whale's back. However, the African or savanna elephant is the largest land animal alive today and indeed has no enemies apart from man in its grassland home.

The African landscape is also the home of the world's tallest mammal. The giraffe's head towers some five metres above the ground. The greatest recorded height is 5·8 metres, which is 1·4 metres taller than a London double-decker bus!

The largest bird – the ostrich – is also found on the grasslands of Africa. This flightless bird, or ratite, stands about 2·5 metres tall and on average weighs 120 to 127 kilograms. The heaviest bird that can still fly is the kori bustard of East and South Africa where adult cock birds average about 12 kilograms, although there is a record of one tipping the scales at 18 kilograms.

At the other end of the bird world, the smallest in size and weight are the tiny hummingbirds, nature's helicopters from tropical America. The smallest is actually found only on the island of Cuba. This is the bee hummingbird, the male of which has a body length of 58 millimetres and weighs about 20 grams, which is less than a sphinx moth.

The giants of the reptile world are the notorious man-eating crocodiles and the constrictor snakes. The largest reptile is the estuarine crocodile which ranges from India eastwards through the Malay archipelago to northern Australia. Adult males average 3·6 to 4·3 metres in length and weigh about 450 kilograms, but even older ones can be half as heavy again. The largest on record is 8·2 metres

Above: Some of the record breakers of the animal kingdom are shown here. The blue whale is the largest and heaviest animal in the world; a female measured 33·5 metres in length. On land the tallest is the African giraffe, while the African elephant is the heaviest, up to 10·7 tonnes. The ostrich is the largest bird although it cannot fly, and the largest snake is the anaconda.

Above: The smallest bird is the bee hummingbird, here compared with a bumble bee for size. Adult specimens measure 90 mm long and weigh between 3·5–4·5 grams.

Invertebrates are all those animals that do not possess a backbone. This includes organisms such as amoeba, sponges, starfish, worms, insects, snails, jellyfish and sea squirts. The animals with backbones are called VERTEBRATES.

Jurassic period (195–135 million years ago) saw the rise of the largest dinosaurs.

Lamarck, Jean Baptiste (1744–1829), was a French biologist who first put forward a theory of the inheritance of acquired characters to account for EVOLUTION. He believed that changes in an animal's con-

Jean Baptiste Lamarck

ditions created new needs for it. These new needs lead to new methods of behaviour and involved fresh uses, or non-usage, of existing organs of the body. Lamarck gave the rather poor example of the giraffe to illustrate his point. Lamarck inferred that this animal in trying to feed on foliage that grows high above the ground, stretches its neck and legs in the process. As a result, many generations later the giraffe had developed into an animal with very long legs and neck.

Living fossil is the term given to many modern animals and plants that are recognizable as relatives of prehistoric forms, and have not greatly changed their form for millions of years. The coelacanth is a living fossil in the fish world. This lobe-finned fish was thought

Coelacanth

for a killer crocodile from the Philippines.

There are many exaggerated claims about the longest and heaviest of snakes – the anaconda of tropical South America. Although these serpents are claimed to have been seen over 12 metres long, most adults never grow to more than six metres. However, a snake shot in the Upper Orinoco River, eastern Colombia, in 1944 was measured at 11·3 metres.

Arthropods are limited in size by the fact that their hard external skeleton has to be shed, or moulted, each time they want to increase in size. The heaviest insect is the Goliath beetle, but it only weighs up to 95 grams. Some crabs may weigh up to 18 kilograms.

Certain animals in the animal kingdom still hold records although they are not the largest, heaviest, or smallest. These are the ones that are able to live in extreme habitats, such as at a very high altitude or great depth.

Speed

The popular tale of the hare and the tortoise clearly shows the relative speeds of the two animals. Although the hare moved fast, almost 400 times faster than the slow, plodding tortoise, it lost that particular race through taking a nap in the middle. Animals have evolved different methods of movement and differing speeds to suit their particular needs and environment. In the air, wings have proved to be the perfect equipment. In water, the animal is usually streamlined in shape (like a torpedo), as this design is best suited to dealing with the water's

Below: A chart showing the altitudes and depths that certain members of the animal kingdom can live at. These are compared with the deepest mine shaft, ocean trench and highest mountain.

3 Tortoiseshell butterfly

4 Common toad

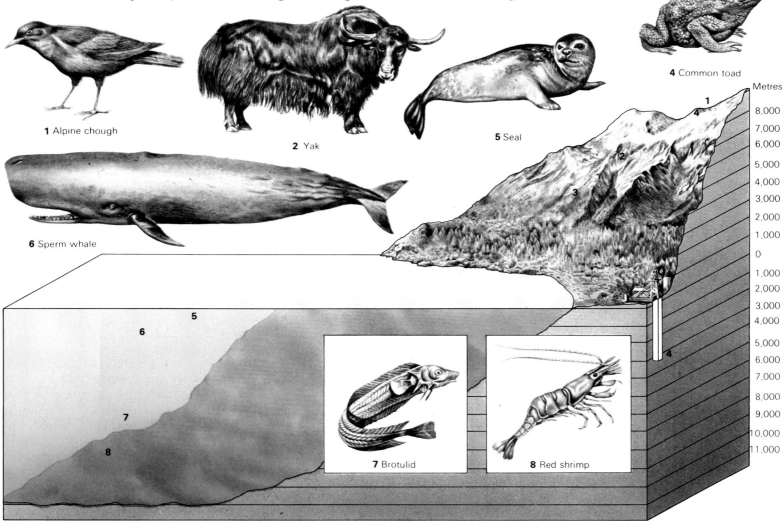

1 Alpine chough

2 Yak

5 Seal

6 Sperm whale

7 Brotulid

8 Red shrimp

Metres
8,000
7,000
6,000
5,000
4,000
3,000
2,000
1,000
0
1,000
2,000
3,000
4,000
5,000
6,000
7,000
8,000
9,000
10,000
11,000

to be extinct for over 65 million years until in 1938, one specimen was caught off South Africa.

M **Mendel,** Gregor (1822–84), was an Austrian priest who discovered a mechanism of inheritance (the transmission of characters from parents to offspring). He performed thousands of experiments with the garden pea to show genetic inheritance.
Mesozoic era is from the TRIASSIC to the end of the CRETACEOUS, and is often called the Age of Reptiles.

N **Natural selection** is the main process by which animal and plant species undergo evolutionary change. Within any given species, there are always variations in the characteristics of individuals. Some of these individuals will have certain characteristics that give them a better chance of survival in a changing environment.

O **Ordovician period** (500–440 million years ago). The first fossil remains of vertebrates have been found in rocks of this period.

P **Palaeontology** is the study of FOSSILS of animals and plants in order to understand what life was like in the Earth's past.
Parasites are organisms living in or on another organism which acts as the host. They obtain food from the host and are usually harmful to it. A tapeworm, for example, lives inside the gut of an animal, and is an internal parasite. A blood-sucking tick attaches itself with powerful jaws to a mammal's body and is therefore an external parasite.
Pre-Cambrian period is the general name given to the first 4,000 million years of the Earth's history.
Pterosaurs are animals be-

Pterosaur

longing to an extinct order of flying reptiles. Their 'wings' consisted of webs of skin stretched between the hind legs and body and the ex-

Gregor Mendel

Far left: Flying fish, here seen in the Red Sea, have large pectoral fins that can be spread like wings. Flights can last for over 30 seconds and the distance travelled can be up to 350 metres.
Left: Ostriches live on the African savanna grasslands. Although they are the largest living birds, they are unable to fly.

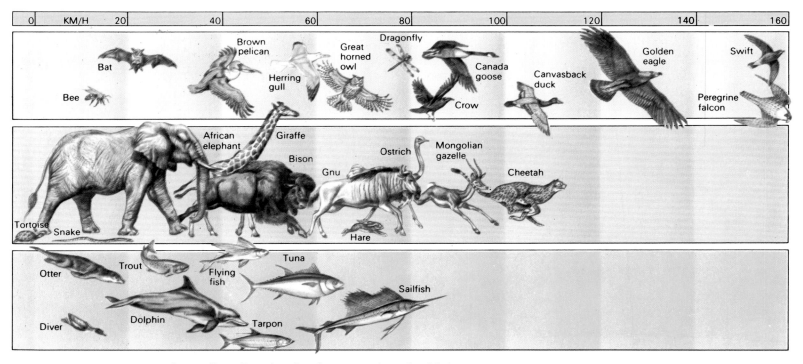

Above: The fastest animals on land, sea and air are here compared. Air offers the least resistance, so the fastest speeds are attained. Streamlining is important in all categories. On land, mammals are the fastest movers, and although some mammalian species, such as dolphins and otters, are fast movers in water, the fishes are the fastest swimmers. Because of the greater resistance offered by water, the speeds reached are much slower than those achieved by animals on land or in the air.

drag effect. Land animals are equipped with legs which can vary in number from just two to over 100 pairs. Certain creatures, such as snakes, are legless, however, and they too have developed their own method of movement.

If all the animals of the world were pitted against one another in a race, the birds would outclass all the others. The swifts are well named as they are among the fastest of birds. A spine-tail swift is claimed to hold the record at 352 kilometres per hour, although this is rather doubtful. This swift has certainly exceeded 161 kilometres per hour so it would still be the fastest.

On land, four legs are generally more useful than two. The cheetah is the fastest animal over a short distance at about 90 kilometres per hour. However, a Mongolian gazelle and an American pronghorn antelope could keep up a speed of

over 80 kilometres per hour for much further. A gazelle, which is being hunted by a cheetah, will often escape if it can jump and turn in the first few hundred metres. The cheetah is not so expert at quick manoeuvres, and after about 350 metres the cheetah will involuntarily pull up absolutely exhausted. The same story is also true of a hare being hunted by a fox, or a mouse attempting to evade a pursuing weasel. In water, quick movement is more difficult because the medium is about 800 times denser than air. The fastest swimmer is probably the sailfish, which can move at over 80 kilometres per hour.

Although speed and movement are important animal features, there are some creatures that do not move at all. The corals, barnacles and some shellfish become fixed to a solid object, such as a rock, early in life and remain there until they die.

Compared with birds or fishes, there are only a few species of mammals. But, since the extinction of the dinosaurs over 60 million years ago, mammals have dominated the land. Other mammals also live in the sea and some are even air-borne.

Mammals

Although relatively few in numbers compared to other groups of animals, the mammals contain the most successful and also the most conspicuous animal forms on our planet. An important feature of this group is that mammals are the only animals that have hair – sometimes called wool or fur – covering their bodies. However, there are some mammals that are relatively hairless. The ELEPHANT, although quite hairy when it is born, loses most of it as it gets older. The sea-dwelling DOLPHINS and WHALES have also lost their fur during the course of evolution.

Hair is a most important factor in maintaining a mammal's body heat, as it forms an insulating (heat-trapping) layer around the body. Mammals and birds are warm-blooded, that is to say they can make heat inside their bodies. This enables them to digest food faster, move faster and grow faster than cold-blooded animals.

Female mammals can also produce milk in mammary glands. These supply their young with important nourishment during early childhood. Pouched mammals, or MARSUPIALS, although giving birth to embryo-like young, suckle their young in a pouch until they are fully developed. The most primitive mammals are the egg-layers, known as the MONOTREMES.

Although there are only 4,200 species of mammals compared with 8,500 birds, or 20,000 bony fishes, the class is a very successful one. Mammals can be found in the air (BATS); in the tops of trees (squirrels, MONKEYS, LEMURS); browsing on the forest floor (DEER, wild pigs); grazing on the grasslands (horses, ANTELOPES,

Edentata: Anteater

Pholidota: Pangolin

Rodentia: Beaver

Cetacea: Dolphin

Hyracoidea: Hyrax

Sirenia: Manatee

Perissodactyla: Tapir

Artiodactyla: Hippopotamus

Monotremata: Platypus

Marsupialia: Koala

Insectivora: Mole

Dermoptera: Colugo

Chiroptera: Bat

Primates: Lemur

Carnivora: Bear

Right: There are 19 orders of mammals. A mammal is an animal that feeds its young with milk secreted by the mother's mammary glands. Most mammals are covered with hair, sometimes called wool or fur.

Pinnipedia: Seal

Lagomorpha: Hare

Proboscidea: Elephant

Tubulidentata: Aardvark

Reference

A **Aardvarks** are the single species of the order Tubulidentata. They live in Africa and have pig-like bodies with a thick, tapering tail and a long tongue which sweeps up termites.
Anteater is the name given to those mammals adapted to feeding on termites and other soft-bodied insects. These include PANGOLINS, aardvarks and ECHIDNAS.
Antelopes are graceful runners of the order ARTIODACTY-

LA. They are found mainly in Africa and the family includes springbok, impala, gerenuk and gazelles.

Apes differ from all other PRIMATES in the absence of a tail, greater brain capacity, longer arms and no cheek-

Tamandua, or Lesser Anteater

pouches. They are man's nearest relatives and include the gibbon, orang-utan, chimpanzee and gorilla.
Armadillos are armour-plated relatives of the SLOTHS and ANTEATERS. They are only found in the Americas and include 9-banded, giant, and fairy armadillos.
Artiodactyla. This even-toed ungulate order includes pigs, camels, deer, antelopes and cattle. They all have 2 or 4 toes on each foot, complex stomachs and the majority have horns or antlers which are placed on the crown.

B **Baboons** are Old World MONKEYS that have adapted to living and hunting on the ground. They live in groups of up to 200 mem-

Baboon

Left: The mole is highly adapted for an underground life with its broad fore-paws that act like highly-efficient shovels, and its tiny eyes. The hind feet kick back the loosened earth.

Right: Many aquatic mammals and birds have webbed feet. This increase in surface area aids movement through the water. Here a European otter shows off his webbed feet.

cattle); high up on mountainsides (GOATS, sheep); living beneath the Earth's surface (MOLES, LEMMINGS); in undergrowth (RODENTS); as active predators (LIONS, BEARS, mongooses) and as aquatic inhabitants (WHALES and SEALS).

Most mammals have evolved a system of movement that uses all four legs on the ground, although there are a few exceptions. Man, for one, has become upright in his stance. His action is known as bi-pedal movement. He only walks or crawls on four legs as a baby. The CHIMPANZEE, GORILLA and ORANG-UTAN are relatives of man and are all grouped in the ape family. They usually move on all fours, but can walk on their hind legs and use their arms to balance as they waddle along. The GIBBON, another of man's relatives, is able to walk on two legs but usually swings from branch to branch using its long arms.

Depending on how fast they need to move, land animals have evolved various different stances. For example, fairly slow-moving animals like BEARS, BABOONS, APES and man, walk on the soles of their feet (plantigrade). In others the sole of the foot is raised, giving greater length to the leg and the animal moves on the digits of the toes and fingers. This method is found in dogs, cats, mongooses, rabbits and rats. The fastest animals have the longest legs in proportion to their body and this is achieved where the mammal moves on tip toes, or the nails or claws of the feet have enlarged and become hooves. The hoofed mammals, or ungulates, all move in this way

(unguligrade), either raised up on one or three toes (Perissodactyls) or two or four toes (Artiodactyls).

Tree-dwelling mammals have grasping hands with five fingers. Often one of the fingers is opposed to the other four forming a thumb. This enables the hand to get a firm grasp of a branch. This is seen in most monkeys, apes, and lower PRIMATES such as tarsiers.

Mammals that burrow under the ground have developed digging equipment. Rabbits, AARDVARKS, ARMADILLOS and prairie dogs have strong thick claws. An armadillo, if pursued, can burrow out of harm's way quicker than a man can dig a hole with a spade. The champion digger is the mole whose front feet are like living shovels.

Several mammals have left the land to spend most of their lives in water. To move efficiently through the water both legs and arms have usually become flipper-shaped as in seals and sealions. In whales and dolphins the fore flippers are present and help with steering, but most of the propulsive force comes from the strong muscular tail.

The webbed feet of the BEAVER and otter help push the animals through the water and the beaver has a splendid paddle-like tail which acts like a rudder in water and a prop on land when the beaver is busy chopping down trees with its strong, gnawing front teeth.

Mammals have also taken to the air. There are over 980 species of bat. This order is second only

bers. They feed on fruit, insects and small animals.

Bats are the only mammals capable of true flight. There are two groups, the fruit bats and the insectivorous or carnivorous bats. Fruit bats use their sensitive eyes whereas the rest use echolocation to find their way about. The few families that live in colder climates either hibernate or migrate.

Bears are heavily-built, practically tailless carnivores with broad flat feet. Most bears are omnivorous in diet, but the polar bear is a true flesh-eater.

Black bear

Beavers are amphibious RODENTS of Eurasia and North America. They have a broad, flat tail which acts as a rudder, and live in family units, building a dam across a stream with a lodge of aspen and willow upstream.

Beaver

Bison, or North American buffalo, and the European bison, or wisent, were almost exterminated by man, but are now protected. The American bison has the shaggier coat of the two.

C Camels are either Bactrian (two-humped) or Arabian (one-humped). Both types have adapted to life in the harsh deserts.

Caribou are North American REINDEER. Both sexes grow antlers. They have adapted to Arctic conditions with thick coats and broad, flat deeply-cleft hoofs.

Carnivora. This order contains a wide variety of mammals including cats, dogs, bears, otters, hyenas, etc. Their jaw-line and tooth structure has adapted for eating flesh, although some will also eat vegetation.

Cat refers to both the various domestic breeds and to wild cats such as the lion, jaguar, leopard and lynx.

Cattle are all ARTIODACTYLS and include both wild species such as the Indian buffalo and YAK and the domestic breeds (shorthorns, jerseys).

Cetacea. This order con-

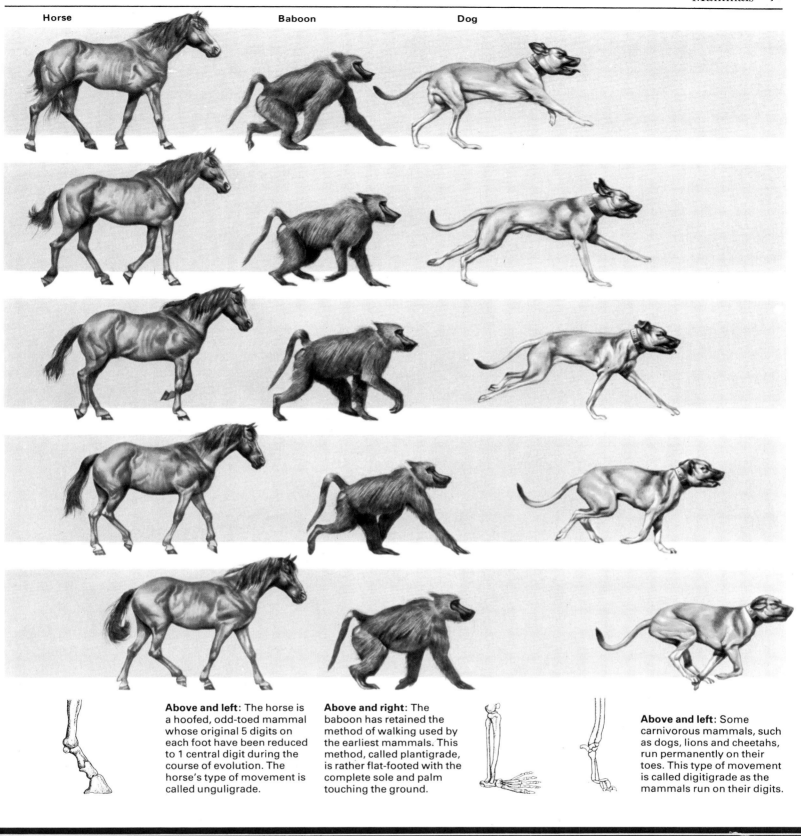

Horse   Baboon   Dog

Above and left: The horse is a hoofed, odd-toed mammal whose original 5 digits on each foot have been reduced to 1 central digit during the course of evolution. The horse's type of movement is called unguligrade.

Above and right: The baboon has retained the method of walking used by the earliest mammals. This method, called plantigrade, is rather flat-footed with the complete sole and palm touching the ground.

Above and left: Some carnivorous mammals, such as dogs, lions and cheetahs, run permanently on their toes. This type of movement is called digitigrade as the mammals run on their digits.

tains the WHALES, DOLPHINS and porpoises. They are all highly adapted for an aquatic life. They use their forelimbs as paddles, have no hind-limbs and their tail is a pair of horizontal fins.
Chimpanzees are APES that inhabit the forests of Africa, living in family groups. They sleep at night in trees and are inoffensive unless molested.
Chiroptera. This order contains the BATS.

D **Deer** are all ARTIODAC-TYLS and the group includes species such as

fallow, red, Virginian and mule deer as well as moose, CARIBOU and reindeer. Most have bony antlers (usually only the males) which are

Young chimpanzee

shed every year.
Dermoptera, see FLYING LEMUR.
Dog refers to either the 130 domesticated breeds or to

Porcupine

the wild species which include the WOLF, FOX and JACKAL of the family Canidae.
Dolphins are apparently highly intelligent CETACEANS. They are often seen in family groups, or 'schools', playing at the ocean's surface and following ships.
Dugongs live in the Red Sea and Indian Ocean and are members of the order Sirenia. Seal-like in shape, they are slow-moving, timid mammals that feed on seaweeds.

E **Echidnas,** or spiny ANT-EATERS, are termite-eating MONOTREMES from Au-

stralia and New Guinea. Their long, beak-like muzzle probes for food. The upper parts of the body are covered with a mixture of spines and hair.
Edentata. This order includes the SLOTHS, ANTEATERS and ARMADILLOS, all of which live in the Americas.
Elephants are the largest living land mammals of the order Proboscidea. There are 2 species, the African and the Indian or Asian. Their trunk is used for feeding, drinking and bathing. The African elephant has the longer tusks and larger ears.

Above and left: Carnivores (the meat-eaters) such as the lionesses illustrated here, have dagger-like canine teeth for piercing and tearing the prey, with slicing cheek teeth that cut the meat into lumps.

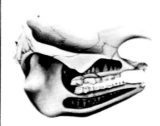

Above and left: Herbivores (the plant eaters) such as the white rhinoceros seen here, have to eat large quantities of plants daily to obtain enough energy for survival. The cheek teeth grind down the plant material.

Above and left: A giant anteater rips open a termite nest using its strong front claws. Then the long, sticky, extensible tongue laps up the insects. No teeth are required for this diet.

in size to the rodents and flourishes in every part of the world except the polar regions. The bat flies using wings that have evolved from the five-fingered forelimbs — webs of naked skin stretching between the digits and usually enveloping the hind legs and part of the tail. Although some mammals are given the term 'flying', such as the FLYING LEMUR and flying squirrel, these mammals actually can only passively glide or parachute on outstretched webbed limbs down to a lower level in the trees.

Feeding and senses of mammals

Mammals, being warm-blooded and with a high energy use, need a certain amount of food in order to keep the body processes functioning correctly. The plant eaters, or herbivores, do not need to use up much energy searching for food. However, winter, drought, or severe climatic changes, may make this food source scarce and so the mammal has to search harder. Another difficulty with this type of diet is that not much energy is obtained from plant material. This means that considerable amounts must be eaten by the mammal to stay alive. An adult African bull elephant, using its trunk to push leaves and branches into its mouth, will consume about 150 kilograms of vegetation a day. Herbivores must have flat, ridged teeth in order to grind down the tough plant material before swallowing.

Certain herbivorous mammals, such as the elephant, TAPIR and giraffe, browse on bushes and trees while many others feed on grasses. The latter species includes BISON, antelopes, white RHINOCEROSES and deer. The gerenuk of Africa is sometimes called the giraffe antelope because it has a long thin neck. If it cannot reach high leaves it will stand on its hind legs to eat.

The members of the RODENT order, some 1,700 species, are basically vegetarians. They all have huge front teeth for biting and nibbling plant matter and powerful grinding cheek teeth.

The CARNIVORES, or flesh-eaters, are those that prey on other animals. These include the large cats, wild dogs, weasels, foxes, and seals. All flesh-eaters have piercing canine teeth, and slicing and cutting cheek teeth. Shrews are also carnivorous but as they eat mainly insects are termed INSECTIVORES.

The fastest land mammal, the cheetah, stealthily tracks its prey, such as a gazelle or antelope,

African elephant

F Flying lemurs, or colugos, are the only species of the order Dermoptera. Cat-sized with large eyes, they do not fly but glide using a membrane which extends round its arms, legs and tail. They range from Indo-China to the Philippines.

Foxes are widespread. Their habitat ranges from the Arctic to the North African deserts. The European red fox has, in recent years, invaded the outskirts of towns to obtain food.

G Gerbils, or sand rats, come from the drier parts of Africa and Asia. Nocturnal, burrowing, seed-eating RODENTS, they are now very popular children's pets.

Gibbons are the smallest APES and come from southeast Asia. Living in family groups, they are expert climbers, hooking their long fingers over branches and swinging along hand over hand.

Giraffes are ARTIODACTYLS found only in Africa. They have long legs and can outrun a horse. Using their elongated neck, they browse on acacia trees in the dry savanna areas south of the Sahara desert.

Gnu, or wildebeeste, are common antelopes of the East African plains. They have a buffalo-like head and horns, with a mane and tail like a horse.

Ibex

Goats. Species in the wild include ibexes and markhors. Wild goats of Europe live on steep and rocky upland areas and are usually extremely wary of man.

Gorillas are the largest APES. They live in family groups in tropical Africa. Inoffensive unless molested, they feed on leaves, vegetables and fruit. Old males attain almost 2 metres in height and 200 kg in weight.

H Hamsters are RODENTS often kept as pets. In western Europe they live in underground tunnels and

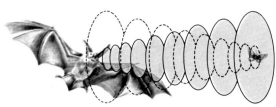

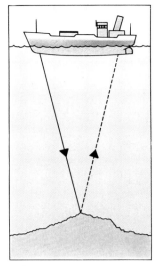

Right: Just as echo-sounding is used by ships to find the depth of the ocean, dolphins and most bats have evolved a similar mechanism to find their way about and detect prey. Short pulses of sound, far beyond the limit of human hearing, are emitted by the mammal. These bounce off objects such as insects or fishes, creating echoes. The brain can interpret the echoes to locate prey or avoid obstacles.

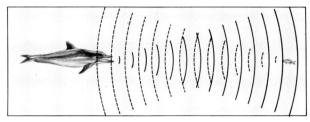

Above: Forwards facing eyes give overlapping, or binocular, vision. The animal sees 2 images of the same subject and can focus well. Thus a cat can judge the distance to jump when moving about from branch to branch.

and then attacks swiftly, leaping and grabbing its victim when within reach. It kills quickly by biting the victim's neck so that the vital jugular artery that goes to the brain is severed. In a pride of lions, the females often hunt in a group, ambushing prey such as ZEBRA, gnu or antelope.

The carnivores of the sea are the seals, sea-lions and certain CETACEANS such as dolphins and killer whales. Depending on the mammal, the prey ranges from shrimps and crabs to penguins and large fishes.

The mixed feeders, or omnivores, are a successful group because they will eat almost anything they can find. The brown bear will eat meat such as salmon or deer when it is available. However, it will happily consume large amounts

Above: A tiger marks his territory by spraying scent from an anal gland over shrubs along his invisible territorial boundary. The scent warns other tigers that they are entering someone else's territory.

of fruit, insects, roots, and grasses in its diet.

If a mammal is preyed upon by others it usually has large ears and eyes positioned on the side of the head so that it can hear and see predators before they get too close. A giraffe has the added advantage of being very tall and grassland animals, such as antelopes, gnu and gazelles, will often will be found grazing nearby. They can thus benefit from their living lookout tower.

Some mammals have eyes facing forwards so that the vision in each eye overlaps to give binocular vision. They are usually either hunters or tree-dwellers. This type of vision allows the animal to judge distances so that, for example, it helps a CHIMPANZEE to swing between branches.

Smells and scents are very important in mammals to bring the sexes together for mating, or to keep unwanted visitors away. Many mammals have large, moist noses (rhinaria) to detect other scents in the air. The skunk has special scent glands near its tail which squirt out an evil-smelling fluid over the advancing enemy, allowing the skunk to escape during the confusion. Other animals such as wild cats, dogs, musk deer, red PANDAS and HIPPOPOTAMUSES use scents to mark out their territories.

A very special type of hearing is called echolocation. This system works rather like radar. The mammal emits high-frequency noises which are bounced back to it off objects in its path. BATS echolocate in the air and dolphins find their way underwater by this method.

---

collect plant food in their cheek pouches. The golden hamster is smaller than the common hamster. The golden hamsters which we now have as pets, all originate from 1 female and 12 young caught in Syria in 1930.

Hippopotamuses are large ARTIODACTYLS distantly related to pigs. They live a semi-aquatic life in the rivers of Africa. They feed on land during the night, returning to the cooler water by day.

Insectivora. This order contains some 285

species of small, primitive mammals. Most members, such as the shrews and tenrecs, have long snouts and feed on insects.

Hedgehog

J Jackals are members of the dog family and find much of their food by scavenging. Most species are a little larger than foxes

and are found in Africa, south-east Europe and Asia.

Jaguars are large cats from South America about the same size as LEOPARDS but heavier in build. They also have a spotted coat, though the spots are larger and have dark centres. Their favourite prey is capybara, a large rodent.

K Kangaroos live only in Australia and are MARSU-PIALS. Great grey and red kangaroos attain up to 2 metres in height and weigh up to 70 kg. They are capable of 40 km per hour and leaps

of 8 metres are not uncommon.

Koalas are small, bear-like MARSUPIALS with tufted ears and a prominent snout. Although looking like teddy bears, they have aggressive natures. They feed only on the leaves of eucalyptus (gum) trees and are now protected species.

L Lagomorpha. This order contains the rabbits, hares and pikas.

Lemurs are small rare PRIMATES that live on the island of Madagascar. They have a fox-like muzzle, very big

shown by other members of the family.

The period of development of a mammal inside the mother's body is called the gestation period. This length of time varies with the species, the larger animals generally having the longest development periods. An elephant is born after about 22 months. The young of the pouched mammals (MARSUPIALS) are born after a very short gestation period. The American OPPOSSUM young are born after 12 or 13 days' development and at birth are no bigger than a honeybee. Although from eight to 18 young are born, less than seven survive the period of further development in their mother's pouch. They remain there for 60 to 70 days attached to, and sucking from, the nipples and then begin to move about her body and accept solid food. The KOALA and the red KANGAROO both have a gestation period of about a month.

The egg-laying MONOTREMES have no nipples.

Left: A red fox vixen offers warmth, protection and milk to her young cubs. At birth these babies are quite helpless.

Below: In the pouched mammals, like this koala, the young are born early but are strong enough to crawl into the mother's pouch.

Reproduction and growth

During the breeding season males and females use various methods of courtship prior to mating and producing a family. Quite often the males live separate lives to the females, but during an annual breeding season they establish a breeding range or territory. For example, the red DEER stag marks out his area using a scent gland, gathers several hinds into his herd and keeps a watch out for any intruding males. If another stag enters his territory, the occupying stag will attempt to see him off with loud roars or actual combat with his fine set of antlers. These antlers are grown during the earlier months and lost after the rutting season (when the deer are sexually active).

The mammal with the largest harem is the fur seal bull, who may tend and guard more than 100 cows, although the number is usually between 40 and 60.

Some mammals live as family groups, such as the LIONS, gorillas and baboons. This has the advantage that a partner is always available for the mating season. When the baby is born, although the mother gives the youngster milk, warmth, care and affection, parental care is also

eyes and a long tail (except for the indri).
Lemmings are furry, stocky rodents of the tundra regions of the Northern Hemisphere. Every 4-5 years, when

Jaguar

there are too many animals for the food available, they migrate in vast numbers. The majority die on the way.
Leopards are large wild spotted cats that range from Africa east to China, with black varieties (black panthers) being frequently found. Excellent climbers, they usually drag prey up into a tree 'larder' where it is safe from other predators.
Lions are mainly found in Africa but a few live in the Gir forest of India. The male has a shaggy mane and heads the family group known as a pride. The

females will often ambush their prey at a waterhole; zebra and antelope being favoured.
Llamas are the South American relatives of CAMELS, but are humpless, smaller, and very woolly. Found only in a domestic state, like the alpaca, they are bred for their fine fur.

M **Marsupialia.** This order contains all the pouched mammals (e.g. KANGAROOS) of Australia and Central and South America. Embryo-like young are born after a short gestation period

and continue their development in the mother's pouch until fully formed.
Moles are underground burrowers of the order INSECTIVORA. Their forelimbs have large digging claws. The ears and eyes are small, the body cylindrical, and the fur is usually black.
Monkeys are agile PRIMATES that are highly adapted to a tree-dwelling life, except the ground-dwelling BABOONS. Most have gripping hands, binocular vision and tails which act as balancing organs. New World species have prehensile tails.

Monotremata. This order contains the egg-laying mammals (duck-billed platypus and the ECHIDNA).

O **Opossums** are marsupials found in America and Australia. The American common opossum is cat-sized and rather rat-like in appearance. They are nocturnal and expert climbers.
Orang-utans are slow, heavily-built primates from the jungles of Sumatra and Borneo. They have naked faces with long, reddish hair over the rest of body. The old males grow large swel-

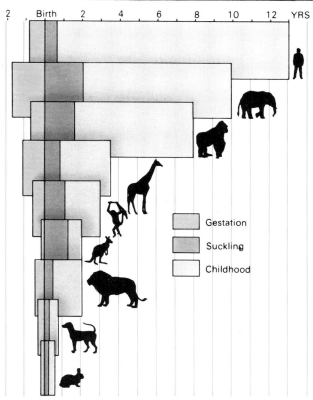

2	Birth	2	4	6	8	10	12	YRS

Gestation

Suckling

Childhood

When a duck-billed platypus has laid her two tiny eggs in an underground nest she incubates them for about two weeks. On hatching, the naked, blind babies feed on the milk which comes onto the mother's belly from mammary pores.

Most mammals give birth to fully developed young, but they are smaller than the adult mammals and rather helpless. These are the 'placental mammals', and include cats, dogs, horses, antelopes and bears. The mother feeds her offspring on her nourishing milk and the amount of care and attention depends on the youngster's development.

GERBILS, kittens and fox cubs are just some of the young that are born blind, naked and helpless. The mother takes great care of them, feeding, cleaning, protecting and keeping them warm. Sometimes she brings her offspring up entirely alone, the father having gone back to living a 'bachelor' life.

Some mammals are able to walk and look after themselves almost as soon as they are born. For instance, a young antelope or gazelle born on the African plain can usually stand and walk within half an hour of birth. Like all mammals it

Above left: The chart compares rates at which different mammals mature and shows the relative duration of gestation (the period within the mother's body), suckling (the length of time until weaning), and childhood (the time taken to achieve sexual maturity). In general, the larger the animal the longer the gestation period, except in marsupials, such as the kangaroo, where the baby is born at an early stage in development.

Above right: A 12-day-old rabbit leaves its warren for the first time. Although blind and naked at birth, its development is rapid and it soon leads an independent life. Other mammals have much longer periods of growing up.

instinctively knows where to find its mother's milk. After feeding it moves off with its mother to rejoin the herd, and it can keep up with the moving herd within a few hours of birth. This is most important, as hunting hyenas, JACKALS, or lions are always on the lookout for strays.

Social life

Within a family group, such as a pride of lions or baboon troop, there are various rules which all the members follow. Usually, each group has a leader. This is a large lion in a pride, or a fit, aggressive, male baboon in a troop. In a herd of elephants, however, it is quite often an old female, as adult bulls are not allowed within certain elephant herds except during the breeding season. Below the leader every member learns his place so that a hierarchy is set up. In a baboon troop, social grooming is an act of behaviour that not only is a form of hygiene but maintains friendly social relations within the group.

Certain mammals migrate on a large scale but the numbers are few compared with the hundreds of species of birds that migrate annually.

---

lings on the face and throat.

P Pandas. There are 2 species. The large, bear-like giant panda of the bamboo forests of China and the cat-like red panda of the mountain forests of western China and the eastern Himalayas. Both are related to RACCOONS. The giant panda feeds almost entirely on bamboo.

Pangolins, see PHOLIDOTA.

Perissodactyla. This order of odd-toed ungulates includes horses, RHINOCEROSES and TAPIRS. They are 1- or 3-toed.

Pholidota. This order contains the African and Asian pangolins. They are toothless ANTEATERS with the upper body and tail covered with horny, overlapping scales, and large digging claws on their forefeet. They roll into a ball for defence.

Pinnipedia. This order contains the SEALS and walruses. Their flippers are limbs modified for an aquatic life. On land they move clumsily but they only come ashore once a year to breed.

Porcupines are large rodents found in North America, Africa and Asia. Their body hairs have evolved into sharp quills which give an effective defence against enemies.

Primates. This order contains tree shrews, lemurs, MONKEYS, APES and man. The brain is usually well-developed and the limbs are usually long with 5 fingers or toes. The thumb and big toe (except in man) are opposable for grasping. Social life is well developed.

Proboscidea, see ELEPHANTS.

R Racoons are nocturnal American mammals recognized by their black and white ringed tail and black 'face mask'. Vegetables and aquatic animals are 'washed' with the forefeet before being eaten.

Racoon

Reindeer is another name for the CARIBOU but usually refers to Eurasian species.

Rhinoceroses are large, thick-skinned PERISSODACTYLS that are inoffensive and retiring in spite of their appearance. They are largely solitary creatures. Five species exist: the Indian, Javan, Sumatran, white and black.

Rodentia. This order contains gnawing animals such as rats, mice, porcupines, beavers and voles. All have a single pair of chisel-like incisors, which grow continually but are kept in check by continual use.

Above: A herd of wildebeeste are here seen crossing a wide river in Tanzania on their annual migration in search of fresh grazing and water.
Right: A dormouse has retired to its sleeping quarters by mid-October. Curled up into a ball to conserve heat, its body metabolism slows down until the warmer weather arrives.
Below: Hamadryas baboons here indulge in a mutual grooming session. This activity helps to strengthen social bonds between various members of the primate troop.

Like birds, mammals migrate mainly to find a better climate, fresh food supplies or to their annual breeding grounds. The migratory instinct is in most cases a combination of these reasons.

Land mammals face many barriers on their journeys — mountains, deserts and rivers being the major obstacles. In Africa up until the present century, antelopes, gazelles, gnu and elephants used to make annual journeys in herds of sometimes many thousands of animals. As a result of hunting by man, these magnificent sights have almost vanished. Gnu still make annual cycles following regular trails in search of grass and water as the dry season progresses. In North America the caribou still migrate in a regimented procession, travelling along established routes to destinations some 1,300 kilometres from their summer feeding and breeding grounds on the high barren ranges of the Arctic tundra. They usually move at six or seven kilometres an hour until they reach the wooded areas of the taiga (a region between the tundra and steppe), where they feed on lichens and the buds and shoots of trees. The caribou of northern Siberia (known there as reindeer) also migrate but they live today in a semi-domesticated state. Man has adopted a nomadic lifestyle in order to follow the mammal that gives him milk, flesh and hide.

There are many species of bat that make seasonal migrations. They have few barriers to face, except for bad weather and wide stretches of sea. In the oceans, whales travel thousands of kilometres annually. The whalebone whales feed mainly on plankton, especially krill (small shrimp-like crustaceans), whereas the toothed whales feed on fish. The summer is spent feeding on shoals of these plankton and fish in the cold waters of the Antarctic. In winter they migrate to warm tropical waters where the young are born.

Some mammals do not migrate when the frosts and snows of winter begin and food becomes very scarce. Instead, they feed almost continuously through the summer months, putting on thick layers of insulating and energy-giving fat. Then they find a suitable dry, draught-free spot, and curl up and sleep until the warmth of spring awakes them. This is known as hibernation and dormice, some bats, hedgehogs, ground squirrels and HAMSTERS are all mammals that go into a deep sleep during the winter.

S **Seals** are aquatic mammals of the order PINNIPEDIA, mainly living in northern waters and in Antarctica. Seals come ashore for breeding, with up to 50 cows forming a harem ruled over by an old bull.
Sirenia. This order contains the manatees, and DUGONGS (sea-cows).
Sloths are very slow members of the order EDENTATA. They live in trees where they hang upside down from branches using their long curved claws. They mainly feed on leaves of the Cecropia tree in South America.

T **Tapirs** are donkey-sized, stoutly-built PERISSODACTYLS found in Malaya and South America. The snout is extended into a short trunk which is used for browsing on foliage.
Tigers are large cats found from India to China and Malaya. The striped reddish and black coat helps to camouflage them in their jungle home. Their diet varies from small game to large animals such as water buffalo. When food is short, or they are injured, they may become man-eaters.
Tubulidentata, see AARDVARK.

W **Wallabies** are small Australian KANGAROOS. Many were formerly hunted for their valuable pelts.
Whales belong to the order CETACEA. They are divided into the toothless whales (blue, humpback) that feed on plankton and the toothed species (DOLPHINS, porpoises) that feed mainly on fish.
Wolves are wild dogs of the order CARNIVORA that hunt in packs.

Z **Zebras** are striped horses of the order PERISSODACTYLA. Three species (mountain, common, and Grévy's) are all found in Africa south of the Sahara. Herds will visit waterholes at dawn and dusk, ever watchful for prowling lions or leopards.

Burchell's zebra

The ancestors of birds were toothed reptiles, but their modern descendants are fully adapted to life in the air, with light bones and streamlined bodies. The annual migrations of some birds are still not fully understood by man.

Birds

Birds are more numerous than mammals, the only other warm-blooded group of animals living today. There are over 8,500 species of birds in existence, but only 4,200 species of mammals. Their success is due mainly to the fact that they can fly. Most species have retained the use of flight but several (PENGUINS, OSTRICHES) have adapted to other methods.

The main identifying feature of a bird is its feathers. Feathers evolved from the scales of reptiles and are the key to their success. They grow from the bird's skin as hair or fingernails grow on humans. However, feathers grow to a definite size and then stop. Although they are firmly fixed in the skin, they suffer from wear and tear, so they are renewed every so often by a process called moulting. Old feathers are usually shed once a year, with new ones growing in their place. This usually happens immediately after the breeding season.

Feathers are vital for many reasons. They trap an insulating layer of air next to the body so that the high body temperature remains constant. Many feathers are brilliantly coloured and play an important part in the bird's social life. The feathers also help to give the bird its streamlined shape and the long, strong tail and wing feathers provide the flight surfaces.

The skeleton of a bird is as light as possible. The bones are hollow and some are fused together to give greater strength. If a bird had to carry its young within its body, then there would be times when it would be too heavy to fly. So the birds retain the egg-laying habits of their reptilian ancestors. An egg is laid usually within 24 hours of fertilization.

The legs of birds are adapted for walking or swimming, depending on the species, and are also important shock absorbers for when the bird lands. To maintain the centre of gravity, the thighs are held close to the body, the knees

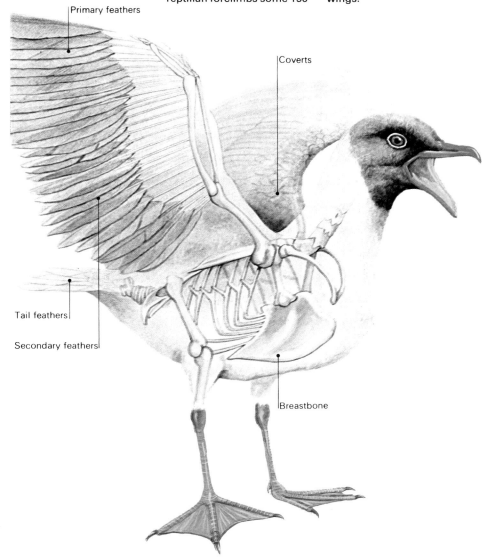

Below: Feathers are unique to birds. Their major functions are to help in flight and provide body insulation. The wings evolved from reptilian forelimbs some 150 million years ago. Massive pectoral muscles are attached to the deep breastbone. These provide the power to move the wings.

Primary feathers

Coverts

Tail feathers

Secondary feathers

Breastbone

Reference

A **Albatrosses** are graceful oceanic sea birds. Their very long narrow wings make them excellent fliers and gliders. They land on oceanic islands only to breed.
Anhingas, or snakebirds, are slender long-necked birds that spear their prey. They throw their victim into the air and catch it so that the head is swallowed first.
Avocets are black and white slender waders with a long up-curved bill which is used to find molluscs, crustaceans and insects in shallow water.

B **Bee-eaters** are brightly-coloured birds of temperate and tropical parts of Eurasia, Africa and Australia. Some species rub the dead bee against a branch to remove its sting. Most nest in colonies in sand banks.
Birds of paradise are found in Australasia. The males are very brightly coloured and are famous for their displays where they dance, pose and hang upside down.

Birds of prey is a term given to all birds, not necessarily related, that are flesh-eaters. They include OWLS, EAGLES, and VULTURES.

Avocet

Bird of paradise

Blackbirds are members of the THRUSH family. Males have a golden bill with dull black plumage. Females are dark brown with a dull orange bill. The name is also given to American birds of the troupial family.
Bobolinks are American blackbirds famous for their migrations. They leave their lush summer northern farmland homes to winter on the pampas of Argentina.
Bowerbirds are less gaudy relatives of the BIRDS OF PARADISE and live in Australia and New Guinea. The males build strange display struc-

Right: The sulphur-crested cockatoo propels itself by the large primary feathers controlled by the 'hand' part of each wing. The secondary feathers and the inner part of the wing maintain lift, while the downstroke is the power stroke. Then the feathers are closed flat so as to encounter the maximum amount of air resistance.

usually hidden beneath feathers. The leg's visible joint is actually the ankle, the bird being raised up on its toes. The majority of birds, such as the perching birds or passerines, have four toes, one pointing backwards and three forwards.

How birds fly

The ability to fly has caused most birds to be designed on a general uniform body plan. However, there is an enormous range in size, from the tiny HUMMINGBIRDS to the largest fliers, the kori bustards. The even bigger EMUS, CASSOWARIES and OSTRICHES are flightless.

The wings of a bird have evolved from the front limbs of its reptilian ancestors. They are streamlined to cut through the air with very little resistance. The shape of the wings vary, depending on the lifestyle of the bird.

Certain birds rely on winds and air currents to soar and glide for long periods. Birds that soar over the land include the VULTURES, EAGLES, kites and HAWKS. They have long, broad wings and can go very high using thermal currents (hot air pockets). Over the ocean the finest soarers are the ALBATROSSES, FRIGATE BIRDS, shearwaters and GULLS. They have light, long, narrow wings that enable them to glide for long periods without flapping their wings. They steer using their tail feathers, and tip their wings and body to bank in turns like an aircraft.

In flapping flight the power comes mainly from the downstroke and from the primary feathers which grow at the tips of the wings. They act on rather the same principle as propellers of a propeller-driven aeroplane. The sequence can be seen in the frame by frame illustrations at the top of these pages. Strong flapping flight is seen in birds, such as GEESE, HERONS, STORKS, CORMORANTS, THRUSHES and most FINCHES.

The aerial acrobats are the tiny humming-

tures which they decorate with flowers. They dance in these 'bowers' to attract a mate. When mated, the female goes off to lay her eggs and raise the young alone.

Buzzards are some 26 species of birds grouped in the HAWK family. All are hunters, feeding largely on reptiles, amphibians and small mammals.

C Canaries are small domesticated FINCHES that originally lived wild in the Canary Islands.

Cassowaries are timid, large flightless birds of the Australasian tropical forests. If attacked they kick with their long legs and sharp claws. A bony helmet, or casque, protects their head as they run through the forest.

Chickens are the domesticated form of the wild jungle fowl from southern Asia. By selective breeding a great variety have been produced. Leghorns, Andalusian and Sussex are just some of the breeds.

Cockatoos are members of the PARROT family. They differ from other parrots in having a crest of long, pointed feathers that they can raise and lower at will. They are found in Australasia where they live in noisy flocks.

Condors are very large American vultures. They have a thick, hooked bill and a bare head and neck. The Californian condor is today one of the rarest birds.

Cormorants are quite large black sea birds with long bodies. They have large webbed feet and a long neck with a slightly hooked bill. They dive underwater to catch fish and can remain submerged for up to a minute. The Japanese train cormorants to catch fish.

Crossbills are small seed-eating FINCHES. The birds' powerful beak can exert 100lb/in^2 at the cutting edge and can cut through the hard outer case of coniferous seed cones with ease.

Crows are bold, noisy and aggressive birds. Their close

Cormorant

Crossbill

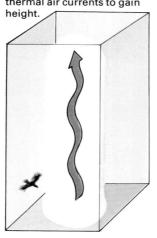

Left: A barn owl returning with a field mouse. The soft plumage helps to make its flight virtually silent.

Above: A hummingbird can fly in almost any direction. Its most remarkable feat is that of hovering.

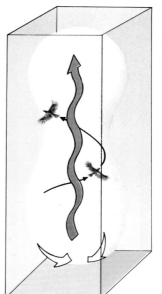
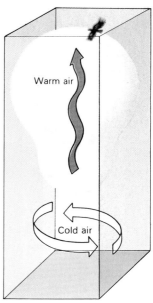

Below: Warm air rises from the hot land and expands. Cold air rushes in and creates a rising hot air bubble. A vulture uses these thermal air currents to gain height.

Warm air

Cold air

birds. They can fly backwards as well as forwards, vertically and stop abruptly to hover. The secret of this bird's success is that the wing is held almost rigid and revolves on a swivel joint at the shoulder. The wing moves swiftly back and forth instead of whirling around like the rotary blades of a helicopter.

Some birds combine flapping and gliding flight. Ibises flap their wings a few times then glide a little farther before flapping once more. WOODPECKERS also do this and their flight pattern is a wavy line. The hummingbirds have the fastest wingbeats per second (over 80) of any bird. They are also fast fliers, up to 95 kilometres per hour by the ruby-throated hummingbird.

Certain other birds can also hover and the technique is usually used to spy for food. A NIGHTJAR will hover for just a few seconds, while a KESTREL has perfected this technique and will hover above a grass verge or meadow waiting for a small rodent to emerge from cover.

Below: An albatross has a wingspan of over 3.5 metres. Its long, narrow wings are adapted for gliding flight over the southern oceans.

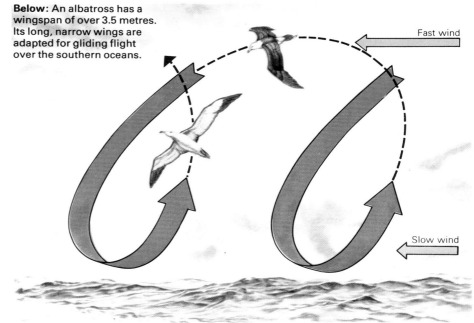

Fast wind

Slow wind

relatives are the ravens, rooks, jackdaws and the more colourful jays.

Cuckoos are famous for being parasitic birds. The female lays her eggs in other birds' nests. She is even capable of producing an egg which matches those of the host species. On hatching, the chick usually pushes any of the host's eggs or nestlings out of the nest. It thus enjoys all the attention of its foster parents. Most species are migratory.

Curlews are brown birds with long legs and a long, curved bill. Curlews often frequent dry uplands, feeding on berries, seeds and insects. They also probe coastal marshes and mud flats for worms.

Golden eagle

E

Eagles are large hawks, and are fine fliers and soarers. They prey on small mammals, birds and amphibians. Nests, called eyries, are often found high up on cliffs. Their hooked bills and huge talons tear their victims to pieces. The golden eagle is perhaps the best known and is found in mountainous areas across Eurasia and North America.

Emus are large flightless birds, related to the CASSOWARY. They are widespread in savannah areas of Australia, feeding mainly on vegetable matter.

F

Falcons are usually solitary birds of prey. Their flight pattern is direct and swift, live prey usually being caught with the talons.

Finches are small, tree-loving, basically seed-eating birds. They are found almost worldwide.

Flamingoes are large water birds that stand over a metre high. They have very long legs, a long neck, and a unique filter-feeding bill. The plumage varies from pale to deep pink. They breed in colonies, each pair constructing a mud nest.

Flycatchers are a large insect-eating family of perching birds. The 350 species, usually called tyrant flycatchers, hunt in the open, catching insects in flight.

Frigate birds are about a metre long and have the largest wing area, in proportion to body weight, of any bird. The wingspan can be up to 2·4 metres. Their plumage is black, the male having a scarlet inflatable throat sac.

Frogmouths are small grey-brown patterned birds that range from India to Australia. The short, broad flat bill has a hooked tip and

Feet and beaks for food

A bird's beak enables it to eat, defend itself, build a nest and preen its feathers. However, the shape of the beak, or bill, is usually designed to deal with the type of food eaten by the species and to enable the bird to reach its particular type of food. The fact that birds, particularly finches, have bills adapted to suit the food available to them was first noted by Darwin in 1835 on the Galapagos Islands.

Seed-eaters, such as SPARROWS, CHICKENS and certain finches, have cone-shaped bills. The sharp point picks up the seeds and the rest of the bill crushes them. A crossbill is well-named as its scissors-like beak can open pine cones.

Below: The frogmouth's wide gaping beak has sensitive bristles that enable it to trap insects on the wing during the night.

Below: The mallard's broad bill has fringes which sift out tiny animals and plants from the water.

Below: The macaw's hooked bill is a powerful and efficient nutcracker. After the case is cracked the tongue extracts the fruit.

Left: A sulphur-breasted toucan displays its huge, colourful beak. The beak is **Below:** The down-curved, long, slim bill of the Kiwi is used to probe the ground for worms. The nostrils at its tip also help to locate food.

made lighter by being honey combed with air chambers.
Below: The bent bill of the flamingo filters out mud and water, but retains the minute plants and animals for food.

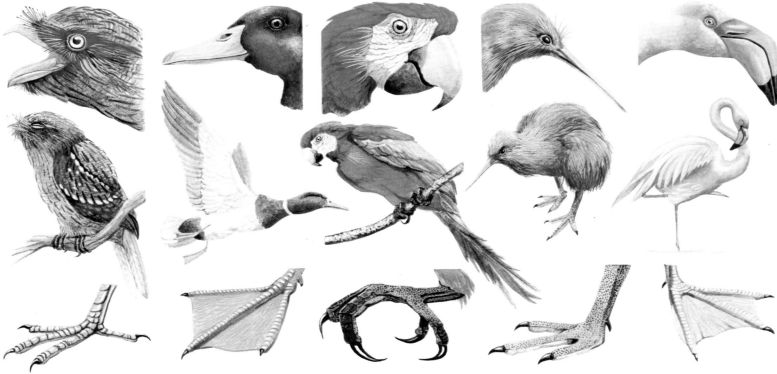

Above: The frogmouth's feet are designed for perching on a branch. Its body plumage usually blends in perfectly with the surroundings.

Above: The webbing between the mallard's front 3 toes gives the duck's foot propulsion through the water.

Above: To help with climbing trees and gripping, the macaw has 2 toes pointing forwards and 2 backwards.

Above: In this flightless bird, the hind toe is greatly reduced. The feet are sturdily built for running.

Above: The flamingo's webbed feet are adapted for wading in mud without sinking, and also for propulsion when swimming.

opens to a huge gape — hence the name. They catch insects in flight.

G Geese are large aquatic birds of the duck family. The best known North American species is the Canada goose, with its black neck and head.
Grebes are long, sleek water birds found almost worldwide. They are famous for their courtship displays. They usually ride high in the water with their long, thin neck outstretched.
Grouse are game birds of the order Galliformes. The

largest is the capercaillie of the Eurasian evergreen forests. The metre-long black male extends his tail into a fan when displaying.

Willow grouse

They eat conifer shoots and buds. Ptarmigan are small grouse.
Gulls are long-winged web-footed water birds found throughout the world. Though often spoken of as 'seagulls', they do not venture far out to sea. They will eat almost anything.

H Hawks are some 40 species of sharp-clawed, long-shanked, fast fliers of the FALCON order. They live mainly on small mammals and birds.
Herons are long-legged, long-necked water birds

with broad, rounded wings and fairly short tails. Unlike the STORKS and FLAMINGOES, their head is completely feathered. They also carry their necks kinked into an S-shape when flying and at rest due to the length of their neck vertebrae.
Hummingbirds are tiny, fast-flying American birds. Over 300 species are known. They have slender, pointed bills with fringe-tipped tongues that can be projected beyond the beak tip.

K Kestrels are small FAL-CONS, distributed over all

the continents except Antarctica. They are noted for the ability to hover on gently flapping wings while scanning the ground for small

Heron

Above: This cassowary's head clearly shows its ear and part of its bony casque.

Nightjars, SWIFTS and FROGMOUTHS have tiny beaks but their gaping mouths are enormous and enable them to catch insects in flight. Members of the parrot family have beaks that act as both nutcrackers and fruit spoons. Eagles and other birds of prey have hook-shaped upper bills that tear their victims to pieces.

Ducks are often termed dabblers because they use their broad, flat bills to dabble in the water and mud to find worms and other aquatic animals. Any sand or dirt is sieved out by the serrated tongue and lower bill. An extreme example of a filtering beak is seen in the FLAMINGOES. Birds have evolved long slender bills where they need to poke into things to obtain food. An OYSTERCATCHER probes into mud for worms and even opens mussels and oyster shells to obtain the soft-bodied animal within. The slender beaks of hummingbirds and their extensible tongues are mainly for reaching inside flowers to obtain the sweet nectar. Spear-shaped bills are designed for catching fish and are found in HERONS and KINGFISHERS.

The scaly toes, feet and legs of birds are designed for walking, perching, grasping, snatching, climbing and fighting. Like beaks, there are many different shapes. Swimmers, such as penguins and ducks, have webs between the toes for pushing and steering through the water. Perching birds of the large order Passeriformes all have toes that can curl around a branch. Climbing birds, such as woodpeckers, have toes

with strong, hook-like claws and usually two toes point forwards and two backwards to give better anchorage.

Birds of prey have powerful toes with strong piercing claws (or talons) which grasp the prey. Flightless birds show a reduction in the number of toes. A CASSOWARY has three while an OSTRICH has only two, although one is very large.

Senses
The most important senses of a bird are sight, then sound, with smell, touch and taste being not so well developed. The eyes of most birds are found on the sides of the head and are only slightly moveable in their sockets. Highly flexible necks enhance vision. When a bird such as a blackbird or robin cocks its head, it is not listening but bringing its eye into a position for better vision. An OWL can turn its head almost full circle.

It may come as a surprise to know that birds have ears, although the openings are usually hidden beneath a feathered head. There are no ear flaps, unlike in mammals, as these would impair flight. Their hearing is excellent and studies suggest their response to sounds is about ten times more rapid than that of humans. Experiments have shown that owls can catch prey, such as mice and voles, by sound alone, and plovers and lapwings can hear earthworms underground.

Smell is important to only a few birds. The

Below: The woodcock has eyes that can see through a complete 360°. This makes it less of a vulnerable target for predators as it busily probes the leaf-litter with its long, sensitive bill.

---

mammals or insects. They have even colonized European cities.

Kingfishers are thick-set birds with short necks and

Kingfisher

large bills. Many have crests, especially the American species. The majority of the 80 species are found in the tropics. Most have brilliant plumage, and often no more than a flash of colour is seen as they emerge from their waterside homes.

Kiwis are the smallest flightless birds and live only in New Zealand. They are quite rare and have an exceptional sense of smell.

L Lyrebirds are spectacular birds of east Australia. The male has a 30 cm long tail with lacy quills. He

performs a shimmering courtship display in which his lacy plumes take on the shape of a lyre as they are spread over his back. Inhabiting forests and scrublands, these birds are shy and largely ground-dwelling.

M Macaws are some 18 species of large South American parrots. Most are very colourful and very noisy, feeding on seeds, nuts and fruit. The larger macaws can crack nuts with their strongly hooked beaks.

Mallee fowl of Australia are

the best studied of Megapodes. These birds do not use their own body heat to hatch eggs. They live in the dry eucalyptus or 'mallee' scrub of southern Australia. They excavate a 1-metre-deep and 3-metre-wide hole, and fill it with vegetation and sand. The eggs are laid in early spring and rain starts the vegetation rotting. The parents regulate the heat build-up by altering the cover over the eggs. The male is thought to test the temperature with his tongue. The birds look like turkeys.

N Nightingales, although famous singers, are unimpressive, small russet-brown birds. They are shy inhabitants of woodlands.

Nightingale

Snow goose

Arctic tern

Cuckoo

Cuckoo shrike

Swallow

Arctic warbler

Swift

Bobolink

Golden plover

Wandering albatross

White stork

KIWI has a good sense of smell, the nostrils being placed at the tip of the upper bill and are used to find earthworms.

Migration and behaviour

Being able to fly allows birds to travel thousands of kilometres from summer breeding grounds to winter feeding grounds. How they navigate is still a mystery but the knowledge, on the whole, is inborn, and they are probably guided largely by the Sun, Moon, stars, and maybe even the Earth's magnetism.

Although a bird can easily cross water, and fly over mountains or deserts, a successful journey depends on the bird having sufficient energy in the form of food reserves for the strenuous flight. Some birds, such as ducks, geese, chaffinches and many songbirds, stop frequently for food. Birds try to avoid travelling over too wide an ocean or desert. For example, the white STORKS and other European migrants avoid crossing the widest

Above: Vast distances are covered annually by thousands of millions of migrating birds. They migrate in order to find adequate food supplies and optimum climatic conditions for survival. They seem to have instinctive navigational skills.

Below: Snow geese over South Dakota on their journey south from the Arctic tundra to California.

Nightjars are more often heard than seen, being nocturnal birds. During the day their remarkable speckled camouflage keeps them hidden on the forest floor.

Ospreys, known as the fish hawks in America, feed almost exclusively on fish. Their large toes have long, curved claws, and spiny scales under the toes hold the caught fish.

Ostriches, found south of the Sahara in Africa, are the largest living birds. Males stand 2.4 metres tall and weigh 130 kg. Flightless,

they travel in groups of 10-50 birds, and are often found with antelopes and zebras.

Owls are largely nocturnal birds of prey. Over 130 species are recognized. Their short, very mobile necks enable them to turn their head almost 360°. Soft, fluffy plumage silences their flight, allowing them to catch prey unawares.

Oystercatchers are large, black and white, noisy birds, seen on coastal shores. The long, blunt, red bill is used to prise open oysters and mussels, kill small crabs and probe for worms.

Parakeets are small parrots with long, pointed tails found mainly in the Indo-Malayan region. They eat grain and fruit, and travel in large flocks.

Passeriformes. This huge order contains all the perching birds or passerines (i.e. more than half the known species of birds). They all have four, similar, un-webbed toes. Their young are always naked and help-less on hatching.

Peafowl are often seen in zoos. The common peafowl originates from southern India and Sri Lanka. The

male, or peacock, grows his magnificent long, eyed feathers to display to peahens. The peahen has a speckled brown plumage for

White pelican

camouflage on her ground nest.

Pelicans are instantly rec-ognized by their large beak-pouch that holds two to three times as much food as its stomach, and is used as a scoop to catch fish.

Penguins are flightless, swimming birds of the coasts of the Southern Hemisphere. Their wings are paddle-like and non-folding for 'flying' through the water. They eat mainly fish, squid and crustaceans. Thick black and white plumage covers the whole body. The largest is the Emperor at

Above: The cat display is a preliminary mating display of crested grebes. If both parties are interested, they then take part in the head-shaking ceremony.

Above: In the head-shaking ceremony, the 2 birds face each other with heads lowered threateningly. They then raise them and spread their head crests.

Above: The ghostly penguin display is where both birds dive and then rise up to face one another before sinking back into the water.

Above: The penguin dance follows head-shaking. They both dive to collect weeds. On the surface they sway from side to side paddling water.

Above: Ceremonies begin in mid-winter and continue for weeks and even months, thus keeping the pair together until nesting begins.

stretches of the Mediterranean and either go round the edge, or use the Straits of Gibraltar or the toe of Italy to cross from. However, millions of birds do take the more difficult, strenuous routes. SWALLOWS, CUCKOOS, wagtails and sandmartins all cross the Sahara.

One of the main ways birds express themselves is vocally. Songs, calls and sounds say what they think of their surroundings or animals that are close to them. Sounds vary from aggressive calls warning that a tom cat or bird of prey is approaching, to wonderful courtship songs that many birds perform in order to attract a female.

Birds compete for living space with other members of their own species. Robins, thrushes, and hummingbirds mark their landowner rights by song. A hummingbird flies continually around its small territory chasing away any rivals. Snipe make a drumming noise by vibrating their tail feathers as they plunge down over their territory.

A bird's plumage is also an important way of

Above left: A courting peacock raises his 2-metre-high train high over his back in a shimmering fan studded with iridescent 'eye' markings.

Above right: A male frigate bird displays to a female by inflating his scarlet throat sac and vibrating his outstretched wings.

communicating. Colourful plumage, together with songs and dances, is an important way of attracting a mate. When a male robin sings, it puffs out its red chest to advertise its presence to a female. The BIRDS OF PARADISE are renowned for their colourful displays. As is often the case, it is the male who is the more brightly coloured, the female being a camouflaged mottled brown. This helps to hide her when she is incubating the eggs and caring for the young birds.

In some species of birds, such as penguins, grebes and herons, it is impossible to tell a male from a female. Yet the birds certainly perform wonderful courtship displays and establish a bond between a male and a female prior to nest building and breeding. Male frigate birds have expandable throat pouches which they inflate into huge red balloons to attract the female. Terns and kingfishers give presents of fish to their prospective partners, while courting GREBES give waterweeds to one another.

---

122 cm tall; the smallest is the blue penguin at 40 cm.
Puffins are members of the auk family. The large, parrot-like bills of both sexes grow a bright sheath during the breeding season. They nest in burrows which they dig with their feet. Fish are caught by diving.

R **Rheas,** sometimes called South American ostriches, are smaller than their African cousins, standing about 1·5 metres high. They are shy inhabitants of treeless open countryside where they live in flocks.

Robin

Robins are the small, plump, friendly birds of gardens and farmlands in Europe. The male sings a wistful and musical song to establish his breeding territory.

S **Secretary birds** are long-legged African hawks, so unlike other falcons and hawks that they are placed in a family of their own. They are so-named because of the long plumes on their head, which suggest a bunch of quill pens stuck behind a 'Dickensian' ear. They are well-known snake killers, but also kill small mammals and birds. They hunt mainly on foot.

Shrikes are small, bold, predatory perching birds. Essentially insect eaters, they will also catch small frogs, lizards, rodents and small

Shrike

birds as big as themselves. Their well-known habit of impaling their prey on thorns has earned them the name of 'butcherbirds'.

Sparrows are small, seed-eating birds. The house sparrow, native to Europe, western Asia and northern Africa is the most successful city and town dweller of all birds. In America, sparrows is the name given to various species of finches called buntings in Britain.

Starlings are jaunty birds with strong legs and feet. The 110 species are generally dark-coloured, most of them black with metallic sheens. Starlings are widely distributed and they mass in huge flocks.

Storks are long-necked,

Right: Swallows build open mud nests under the eaves of buildings.

Above: A grebe anchors its floating nest to reeds.

Above: A kingfisher pair excavate their nest-hole in a river bank.

Above: The South American oropendola's hanging nest is woven by the female from leaf fibres and vine stems.

Breeding and rearing

After territories and pairs have been formed, usually in the spring, then a couple can get down to the serious business of building a nest. A bird's nest will serve as a cradle for the eggs and as a temporary home for the young when they hatch. Usually, they will only leave when they can feed themselves and are able to fly. Nest shapes range from the simple cup-shaped nests of BLACKBIRDS to the intricate woven nests of WEAVER BIRDS. Some cliff-nesting birds like cormorants do not build any nest at all.

Birds use all kinds of material for their nest. Swallows use mud pellets. The TAILOR BIRD neatly 'sews' a large leaf together in its tropical Asian home and then makes its nest inside the leaf. Some cave swiftlets in Asia make their nests out of their saliva and it is these that are collected to make the delicacy, bird's nest soup. Male BOWERBIRDS make a courtship bower, often decorated with berries or brightly coloured flowers, where the female is courted and mated. She then goes off to build her nest nearby.

Many birds excavate burrows in the banks of streams or hillsides. These include kingfishers, bee-eaters and sand martins. Woodpeckers, hornbills, owls and parrots often prefer a hole in a tree trunk. The WOODPECKER can excavate a hole but the others rely on unoccupied holes.

The largest nest is built by the megapodes of Australia. A pair of birds construct a huge mound of rotting vegetation as much as three metres high and 16 metres across. The female lays her eggs in this mass and they are kept warm by the heat given off by the rotting vegetation. If the eggs become too hot the adults scratch away some vegetation to cool them off.

King and emperor PENGUINS lay single eggs which the males carry using their wide webbed feet. The egg is thus not touching the freezing ice and is kept warm by folds of feathered skin on the birds' bellies. The females go off to feed and

---

long-legged birds with broad wings and a long bill. They are usually coloured black and white. The majority of the 17 species fly with the neck outstretched and the feet trailing behind. The European white stork is regarded as a good omen.

Swallows are small, slender-bodied birds with very long pointed wings. Some of the 75 species have forked tails. Many species are migratory. Quick and extremely agile in the air, they are stranded if accidentally grounded because of their short limbs and wing width. They feed on insects caught in flight.

Swan and cygnets

Swans are graceful, large water birds related to geese and ducks. They are the largest of the waterfowl, their short legs being used mainly for propulsion. The long neck is used to reach below the surface of the water to obtain the waterweed on which they feed. The mute swan is not silent, as it can growl and hiss.

Swifts, the fastest fliers of all animals, rarely rest. They spend a lot of time on the wing with mouths agape to catch insects. Their small hooked feet enable them to cling to the sides of cliffs.

T **Tailor birds** are ingenious 'weavers' of southeast Asia. They build their nests by sewing together the edges of one or two leaves with plant fibres or silk from insect cocoons. The cavity is lined with fine grasses, hairs and plant down.

Terns, found throughout the world, are members of the GULL family. Usually smaller and more graceful than gulls, they are often called sea swallows.

Thrushes are fine songsters that live everywhere except the polar regions. The bold mistle thrushes are the largest European species. The smaller, darker brown song thrushes are shyer although they have invaded some city gardens.

Tinamous are primitive, partridge-like birds of South America. They are poor fliers and do not migrate. An unusual fact is that the male incubates the eggs.

Toucans are birds of the South American jungles and have huge, brightly-coloured bills. The bill is hollow but well braced and serrated to help deal with its fruit diet. They are related to WOODPECKERS and barbets.

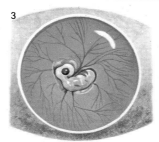

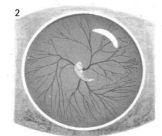

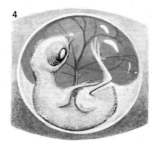

Above: Birth of a chick.

Incubation: The embryo in a new-laid hen's egg is visible as a pale spot on the yolk.

Day 5: The embryo is now surrounded by a network of blood vessels which bring nourishment from the yolk to the developing bird.

Day 12: The head has an eye and a beak, the limbs are beginning to form with the leg having 5 toes (one will disappear).

Day 18: The main flight feathers have developed and only 4 toes with claws are now present. The embryo is fully formed by day 24.

Day 29: The chick has grown rapidly during the last few days and chipped through the egg shell with its egg tooth and struggled free.

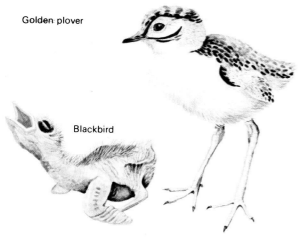

Golden plover

Blackbird

return just in time for hatching. The largest egg laid today is that of an OSTRICH which is about 1.5 kilograms. The smallest is the hummingbird's which weighs only 0.15 grams.

Each egg laid by a mated female bird contains a living, developing bird known as the embryo. It must be kept warm or it will die and so the parent birds, or just the female, sit on the eggs to keep them warm. The embryo develops, living on the yolk until it fills the inside of the egg and is ready to hatch. This development period is called incubation and its length varies considerably. Small singing birds hatch their eggs within a fortnight, while the royal albatross takes 81 days.

On hatching there are two main kinds of birds. There are those that are blind, naked and helpless, and there are those that at birth are fluffy chicks that can see and can leave the nest almost immediately.

Parents of fluffy, well-developed chicks usually lead them away from the nest on the day they

Above left: A young blackbird on hatching is blind and almost naked, and totally dependent on its parents for food and care. However, a golden plover chick is alert and covered with down, being able to leave the nest, run and find food for itself within 3 days.
Above right: A communal penguin crèche in Antarctica. Young penguins are 'in care' for 4 months or more, being looked after alternately by both parents and 'aunties' in the creches.

hatch. The nest is usually on the ground so although the chicks cannot fly, they will not come to harm. Ducks, grebes and geese lead their chicks to water and give them their first swimming lesson. Grebe chicks are often carried on the backs of their parents whilst swimming. Quails, grouse and pheasants take their young on hunting trips and hide them whilst the parents go off for food.

All chicks are fed, protected and guarded by their parents, or just by the female, until they have grown flight feathers and taken to the air. They are now fledglings. Many small singing birds fledge in under two weeks. The emperor penguin chick is fed by its parents for up to 39 weeks before it swims off, while a young wandering albatross may be looked after for up to 45 weeks. A king penguin chick has the longest parental attention. It is 'in care' for 10 to 13 months. Most birds will breed in the first season after they have hatched.

V Vultures play an important role in clearing up dead animals. The American vultures include the CONDOR, while the Afro-Asian vultures are related to hawks and eagles. They usually have a naked head and neck which enables them to keep reasonably clean when feeding on carcases. They soar using hot air thermals.

W Warblers, are small insect-eating birds. Two distinct families are given this name. In the Old World the warblers are relatives of the thrushes, while in the Americas they are relatives of the tanagers. American warblers have brightly-coloured plumage and repetitive songs. The Old World species are rather dull in colour but have long and melodious songs.
Weaver birds of Eurasia and Africa are birds that have brought social development to its highest point in the bird kingdom. They are named after the highly complex communal nests many of them weave. These are elaborate suspended nests woven from vegetable fibres, some of which have entrance tunnels.
Woodpeckers are highly adapted for tree life, with gripping feet and stiff tails that act as props. Their straight, hard pointed, chisel-like bills dig into the tree bark and wood. A long, extensible tongue with a barbed tip extracts insects and grubs. Most species have loud harsh voices and also drum with their bill to advertise their presence.
Wrens are diminutive, busy, brown birds with big voices. They scurry around in the undergrowth searching for insects. The 59 species are wide ranging and have adapted to a variety of habitats.

Y Yellowhammers are small buntings with streaked yellow heads, commonly seen in open country. They are mainly seed eaters.

Vultures

The first land vertebrates were amphibians, part of their lives being spent in water. Their descendants, the reptiles, were the first true land animals, because they did not return to the water to breed.

Reptiles and Amphibians

Right: At almost 3 metres in length, the komodo dragon is the largest living lizard species. It kills and eats animals as big as hogs and small deer. A typical reptilian feature is its scaly skin which prevents the creature from drying out.

Above: When the mouth of an alligator or caiman, is closed, the 4th tooth in the lower jaw fits into a pit in the upper jaw and is hidden. In true crocodiles, the 4th tooth is still visible when the mouth is closed.

Amphibians were the first vertebrates to emerge from water onto the land, towards the end of the Devonian period. Even after some 350 million years they are still tied to their watery element, as most species, in order to breed, must enter water at certain times, and the young begin their development in water. Most adult amphibians have true lungs and can breathe air directly, but they also breathe through their moist skin and through the lining of the mouth which is well-supplied with blood vessels. Because of their moist skin, amphibians must always remain near or in water, or in places with high humidities, such as the steaming jungle. Only three groups of amphibians are found alive today: SALAMANDERS and NEWTS, often termed the Urodelans, the Caecilians, and the FROGS and TOADS which are often termed Anurans.

The reptiles that exist today evolved from amphibians but are a mere remnant of the forms that flourished some 200–70 million years ago during the age of the dinosaurs. One advantage over their amphibian relatives is that a reptile's body is protected from damage and dehydration by its covering of dry, horny scales. As a result reptiles are not tied to water, and can even live in hot deserts. Another advantage is that they do not have to return to the water to breed.

There are four orders of reptiles living today. The Chelonia include the sea TURTLES, land TORTOISES and freshwater TERRAPINS and are identified by their box-like bony shells. The largest reptiles and the closest surviving relations to the dinosaurs are the CROCODILES, GHARIALS,

Reference

A Adders, as they are known in Britain, are the common European VIPERS, although many other species exist. They are venomous, but the bite is rarely fatal to man. The females are longer than the males, being up to about 75 cm. They often have an X-shaped pattern on their head and a dark zigzag along the back.

Agama lizards, of which there are about 300 species, are the counterparts in

Adders

Europe, Asia, Africa and Australia of the South American IGUANAS.

Amphibians are the most primitive class of land-living vertebrates. In most cases, the young are tadpoles with gills for breathing in water. They change gradually into lung or skin breathing land-dwelling adults, although most still return to water for breeding. There are three living orders: the Apoda, Urodela and the Anura.

Anole lizards are some 165 species of IGUANAS living mainly in the Americas. Most species are between 13 and 23 cm long. Their expanded fingers and toes enable them to climb well.

Arrowpoison frogs are small, brightly-coloured frogs of the forests of Central and South America. All have potent poison glands and are so named because they are caught by the Indians and their poison is extracted to be used on arrow tips.

Axolotls are the larval stages of the American salamanders which breed while keeping their gilled larval form. This is known as neoteny and the species occur around Mexico City.

B Bearded lizards are Australian agamids which, when teased, turn

bright yellow and orange. They soon regain their dark olive-brown colour. The beard consists of a great many pointed scales on the throat and neck which the lizard can greatly distend. This ferocious display, especially when its large mouth is opened, usually causes the enemy to hesitate and the lizard can escape.

Blind snakes are some 150 species that have adapted to burrowing and their eyesight ranges from poor to almost non-existent. Mainly tropical species, they feed on worms and millipedes.

Left: The basilisk of tropical America lives on river banks. Quite remarkably, it can rear up and run on its hind legs, using its tail as a balance. In this semi-erect posture it can actually run over the surface of water for short distances.

Left: This snake in the Namib Desert, South-West Africa, is using the sidewinding method to climb a sand dune. A ladder-like succession of furrows track its course.

Right: The chameleon is the sharp-shooter of the reptile world. Its sticky tongue is catapulted out when an insect flies too close.

alligators and CAIMANS of the order Crocodilia. The order Rhynchocephalia contains only one species, the TUATARA from New Zealand. The Squamata contains the LIZARDS and SNAKES which together make up the largest number of modern reptiles, as well as being the youngest group.

Feeding and movement

Most amphibians and reptiles are hunters, and swallow their prey whole. This explains why they have wide gapes to their mouths. They do not usually chew their food so no cheeks are needed. Amphibians feed mainly on invertebrates such as worms, insects and their larvae. Frogs and toads swim in water using their webbed hind feet to give most of the propulsion. On land they hop using their powerful back legs. Newts and salamanders swim by undulating their long body and tail, or walk and scurry on land with a twisted gait. The tongue in frogs and toads is fixed at the front end while the back end lies free.

When an insect comes within range, the tongue is flicked out at high speed and the sticky end collects the prey and returns it to the mouth. Newts and salamanders do not have this type of tongue and so snap at their prey.

In the reptile group, many of the Chelonia, all of which lack teeth, are vegetarians. The carnivorous reptiles, such as snakes and lizards, seize prey by their teeth. The salivary glands in the mouth coat the prey so it is slippery and it can then be swallowed and taken into the digestive system more easily. Chameleons catapult their muscular tongue out to capture small prey.

Snakes move by one of four different methods, or a combination of all four. Wriggling their bodies in S-shaped waves that pass from the head down the length of the body is the most common method. This type of movement relies on the snake's body touching stones, pebbles or plants in its path to push itself forwards. The 'concertina' method is where the snake loops its body

Below: A snake has independently moveable jawbones that allow quite sizeable animals to pass down into its elastic throat.

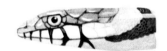

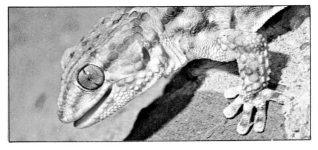

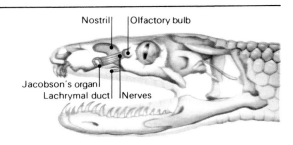

Left: A gecko has cling pads on its feet that catch on any tiny surface irregularities – even glass can be scaled.
Right: Jacobson's organ is the specialized sensor that enables snakes to track prey.
Below: This Texas indigo snake is about to slough its skin. Its flicking tongue picks up particles for analysis by Jacobson's organ.

forwards, the 'rectilinear' method is similar to an earthworm's movement, and 'sidewinding' is used mainly in deserts and sandy places where the snake moves in a series of sideways steps across the surface of the dune.

As a snake moves along it continually flicks out its forked tongue to collect tiny particles from the air and ground. Sensitive cells in the Jacobson's organ in the roof of the mouth detect smells. The reptile can thus track down prey. To overpower its prey a snake uses one of two methods. One method is on having caught the prey with its teeth, the snake then winds its body around until its victim can no longer breathe. These constric-·tors include the BOAS and PYTHONS. Other snakes use poison to overcome their victims. Snake venom usually acts upon the blood and tissues, or upon the nervous system. SEA SNAKES, COBRAS, VIPERS and PIT-VIPERS all have poison glands which are modified salivary glands in the upper jaw. In the upper front row of the teeth are two or more big poison fangs that are visible when the jaws are open. On biting the prey, the fangs work

Below: The elongated snout of this Indian gavial is associated with its almost exclusive diet of fish, caught by a sideways sweep.

like hypodermic syringes. The poison runs through the tubes inside the hollow teeth and is injected into the prey. Prey larger than the snake itself can still be swallowed as the jaws are loosely hinged and the whole region of the mouth, neck and ribs is very elastic, so although swallowing takes quite some time, it is eventually achieved.

All crocodiles have long snouts and tremendous jaws which have peg-like teeth in them. On land they are often seen lazing in the sun. In the water they are almost completely submerged except for the eyes and nostrils and the leathery back breaking the water's surface. When it is close enough to a victim the crocodile moves with a sudden burst of energy, grasps the victim in its jaws or knocks it over with its lashing tail. A large animal such as an antelope or deer is then dragged underwater to drown. The air-breathing channel in the crocodile is completely separated from its mouth so that the reptile can hold and tear and swallow food underwater, yet breathe at the same time.

Cobras are highly poisonous snakes related to mambas, coral snakes and SEA SNAKES. Their short tubular, or grooved, fangs inject poison as the snake chews its victim. The king cobra, or hamadryad, of south-east Asia is one of the largest poisonous snakes and can be up to 5 metres long.
Constrictors are some 76 species of large, non-poisonous snakes that kill their prey by constricting. The victim is encoiled and squeezed until it can no longer breathe and dies of suffocation. Nearly all the

species are tropical, the largest one being the reticulated PYTHON which can measure up to 10 metres.
Crocodiles live mainly in tropical rivers although the estuarine crocodiles do venture out to sea. They are known to attack man but mainly prey on large animals such as cattle. CAIMANS and alligators are included in the same family.

E **Edible frogs** are bright green with yellowish thighs and a dorsal stripe. They are found in Europe and the back legs have been

a favourite item of food since Roman and probably prehistoric times.

F **Frilled lizards** of Australia and New Guinea can run on their hind legs with the tail and forelegs off the ground. They confront their enemies by raising their frill and opening their mouth which makes them look much larger.
Frogs are generally smooth, slimy, shiny-skinned AMPHI-BIANS with very long hind legs that enable them to make very long hops. Their TOAD relatives usually have

dry, rough skins with wart-like growths on them. They have shorter back legs and do not make such long hops. The two names are often interchanged.

Fangs of a puff adder

G **Geckos** are some 300 species of lizards that are found in tropical and sub-tropical regions. The species best known to man is the house gecko which can be seen, usually after dark, running across walls and over ceilings in search of food. They can do this as they have special plates on the under-surface of their toes called lamellae which have adhesive powers. All geckos make soft chirruping or clucking sounds. Tokay geckos have very loud cries.
Gharials are long, thin-snouted crocodiles that live

Right: The female green turtle of tropical waters lays up to 100 eggs in a pit dug in the sand above the high water tidemark. After laying she struggles back to the sea. On hatching, the young turtles instinctively head for the sea but predators such as gulls and crabs take a very high number for food.

ledges the approach of a female by nodding his head, and he makes low pitched roars before mating. Bull alligators roar in the courting season and this is probably done to attract females to them. Male Galapagos IGUANAS and MONITOR LIZARDS have ritual combats between one another. Male RATTLESNAKES, vipers and mambas all entwine their bodies and rear up and push against one another until one glides away exhausted. In some instances a female has been seen resting nearby and has been courted by the victorious male, having overcome his rival and put him to flight.

Eggs are usually deposited in holes dug in the ground. The female turtle lays her eggs in a dry, warm hole in the sand of a tropical beach above the high tide mark. Crocodiles and some alligators are known to build nests and the batch of 20-80 eggs are carefully guarded. The female sits on the nest or lurks nearby. When the young are trying to hatch, the female has been seen to remove the vegetation to help the young escape.

Stories, such as Rudyard Kipling's *Rikki-Tikki-Tavi*, of cobras guarding their eggs are not entirely fictitious. Indian cobras and pythons brood the eggs by coiling round them.

Reptile reproduction

Most reptiles lay soft, leathery eggs. A few give birth to live young, such as the common lizard and CHAMELEON. The developing embryo inside the protective egg case is supplied with a large yolk to nourish it, and a special sac, the allantois, enables oxygen and carbon dioxide to pass between the young and the outside world. The young reptiles, whether alligator, snake or turtle, look like miniature adults and do not have the larval stages of the amphibians.

The sexes look alike in all groups except in certain lizard species. Male agamas and iguanas become brightly coloured and territorial during the breeding season.

Mating is preceded by some kind of courtship in most reptiles. A male giant tortoise acknow-

Above: On laying her eggs, a female carpet python pushes them together into a heap and coils herself around them until they hatch. This may take up to 80 days.

Amphibian reproduction

Most amphibians have to return to freshwater to breed. Their eggs are usually fertilized outside the body. In frogs and toads, the male clings to the female's back, and sprays his semen over the eggs as they emerge from the female. They are laid in huge numbers in jelly-like strings or clumps in most species.

Some frogs take great care of their brood. The male midwife toad carries the eggs around until hatching, wrapped round his hind legs. He spreads the eggs out just as they are about to hatch. The marsupial frog of Brazil keeps the eggs safe in a pocket of skin.

---

in the Indus, Ganges and Brahmaputra rivers in India. This snout is an adaptation for catching fish.

Giant tortoises are 2 huge, long-lived species, one inhabiting the Seychelles and Aldabras, and the other inhabiting the Galapagos Islands. They probably reached these islands via ocean currents, being able to withstand long periods of starvation. They are strictly herbivorous.

Gila monsters, the only venomous lizards, are black and yellow beaded in appearance and live in south-

west North America and Mexico. The poison glands are in the lower jaws.

Greater crested newts are Eurasian AMPHIBIANS. The

Giant tortoise

females are larger than the males, growing up to 18 cm long. Males have a dorsal crest which is absent in the females.

Green turtles are found living in all warm seas, but mainly within the tropics where they are killed to make into genuine turtle soup. This, and the egg-collecting done by man, has led to the species becoming quite rare in places. Certain beaches are now protected for breeding purposes.

H **Horned vipers** are also known as sand vipers. The broad head has the tip of its nose drawn out to form an erect and scaly horn. They live in North Africa, Arabia and south-west Asia.

I **Iguanas** are some 700 species of lizards found mainly in tropical America. They usually live in trees, but are strong swimmers

Iguana

Right: Most frogs mate and lay their eggs (spawn) in water. Jelly surrounds each egg and acts as protection (1). At first the tadpole breathes through feathery gills, but these are lost when internal lungs develop. (2). By 10 weeks the tadpole is frog-like (3) and within a month the tiny amphibian can leave the water (4).

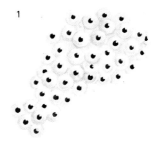
1

2

3

4

In newts and salamanders, the eggs are fertilized within the body of the female and before pairing there is usually an elaborate courtship display. The males often develop bright mating colours and after an aquatic courtship dance, he deposits his packet of sperms, known as a spermatophore, beside the female. She picks it up and stores the sperm until it is needed to fertilize her eggs.

Below: Having spent most of the year in damp habitats on land, in spring the great crested newts enter water to breed. The male develops a fine dorsal crest and both have black and red bellies. The eggs are attached singly to water plants and a leaf is folded over them.

Most amphibian young go through various larval stages. On hatching they are limbless and have external feathery gills and a broad tail. As they develop, legs appear, the external gills are lost when the lungs begin to function and the tadpole begins to resemble a small adult. The tail fin in Anurans is re-absorbed and the small adult is complete and can leave the water, only returning to breed.

and take readily to water. The marine iguana of the Galapagos is the only lizard to live in the sea.

L **Lizards** come in all shapes and sizes from a few centimetres to the 3-metre-long Komodo dragon. The slow-worms and burrowing lizards have lost their limbs and have small eyes and ears. A few lizards, such as the skinks, give birth to live young, but most lay eggs. Many lizards, such as the CHAMELEONS, can change their skin colour to match their surroundings.

M **Monitor lizards** are large, fast-running predators that live throughout the warmer regions of Africa and Australasia. Among the 24 species are the 3-metre-long Komodo dragon and the Nile monitor. The Komodo dragon will eat domesticated animals while most monitors eat amphibians, birds, and eggs.

N **Newts** are a type of SALAMANDER and the 20 species belong to the genus *Triturus*. Most species are found in Europe. They have short legs, a long body and

tail, and a typically moist skin. They are mainly land-dwelling but return to water for the breeding season. During this season, the male develops a crest.

P **Pythons** belong to the BOA CONSTRICTOR family and are found mainly in Africa, Asia and Australasia. The reticulated python can reach a length of over 10 metres. The royal, diamond, and Indian pythons, are amongst the most beautifully patterned snakes.

Pit vipers are so-called because they have a pit in the

side of the face between the eye and the nostril that acts as a sense organ. Rattlesnakes, bush-masters and fer-de-lances are all pit vipers.

R **Rattlesnakes** are so-called because their tails end in a series of horny, interlocking segments. When the tail is vibrated

Eastern diamondback rattlesnake

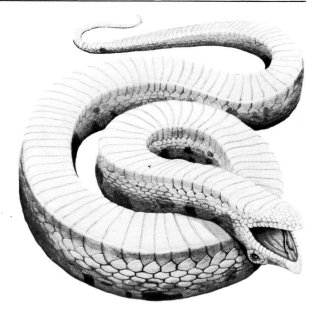

Reptile behaviour

Like fish and amphibians, reptiles are often said to be 'cold-blooded'. As a result they exhibit a common act of behaviour, that of basking in the sun to raise their body temperature. Their body temperature varies with their surroundings. In very cold climates they become lethargic and grass snakes, common lizards and pet tortoises must hibernate through the cold winter months. Reptiles flourish best in tropical conditions where they can be active enough to compete with other creatures.

The wear and tear to the skin of lizards and snakes means that it needs to be replaced at intervals. There are several moulting periods each year. In most lizards, such as the chameleon, the skin is shed in large flakes, while in most snakes it is shed in a single continuous slough. It is possible to tell when a snake is about to moult as the eyes take on a bluish tinge. A snake's eye is covered by fused transparent eyelids and these are replaced in the moult.

Some reptiles have the striking ability to change the colour of their skin. CHAMELEONS are the most accomplished at this and lighting conditions seem to be the most important factor influencing colour change. In complete darkness the animal becomes very pale, and it tends to darken when exposed to light.

Many reptiles have appendages growing from

Above: A chameleon sheds its skin in quite large irregular plates, unlike the single continuous slough in snakes.
Above right: A hog-nosed snake lies inert, belly uppermost and tongue protruding, feigning death. It will come to 'life' when all is clear.
Below: An Australian frilled lizard's warning device is to erect its frill and face its aggressor. This makes it look much larger.

their bodies. There are horned chameleons, and IGUANAS and AGAMA LIZARDS have throat fans. The large frill of the frilled lizard and the spiny 'beard' of the bearded lizard can be raised and lowered rapidly and displayed in aggressive or courtship gestures.

One of the most interesting skin structures is the rattlesnake's rattle. It consists of a number of interlocking horny segments. It is now generally accepted that the rattle warns intruders to keep their distance, although previous theories ranged from a mating to a distress call.

rapidly, it produces a dull hissing sound (rattle) to warn enemies. A new segment is added each time the skin is shed but the end ones continually break off, so 6 to 10 segments is usual.

S **Salamander** is the name applied to many species of families of the order Urodela. The spotted, or common, European salamander lives in hilly districts and hides by day under stones or decaying leaves. It is black with bright yellow spots. The secretions of its skin are poisonous to small animals. The AXOLOTL of Mexico is a salamander. The giant salamander of Japan is the largest known and can grow to a length of about a metre. Its flesh can be eaten.
Sea snakes have successfully adapted to a marine life. Most species have flattened bodies for swimming and give birth to live young. They produce an extremely potent venom.
Skinks are some 700 species of LIZARD that have small limbs, although some are limbless. Half the species lay eggs while the rest give birth to live young.

Snakes are reptiles that are legless, shed their skin completely, have no external ear openings and usually swallow their prey whole. There are about 2800 species. They kill by either injecting their prey with venom or by constriction.

T **Terrapins** are freshwater TORTOISES or TURTLES. Americans call large water turtles, terrapins, while the British sometimes call them water tortoises. They are termed cooters or sliders in the southern United States.
Toads are strictly members of the family Bufonidae. They are more slow moving than FROGS, their heads are more rounded and their skins are warty. Toads lay their eggs in strings as opposed to the cluster-type spawn laid by frogs.

Toads mating

Amphibian behaviour

Amphibians have few defences against their enemies and so most of them rely on protective coloration to camouflage themselves. Most frogs and toads are brown, yellow or green to match their habitat. Tree frogs are usually as green as the leaves on which they live, and so are very difficult to see. Being so well camouflaged, and also quite small, means that at the breeding season sound is very important in getting the sexes together.

A powerful weapon of defence is poison and many amphibians are equipped with poison glands in their skin. In most toads the poison glands are massed into warty lumps so that a predator on picking up a toad in its jaws will quickly drop it. The arrowpoison tree frogs of South America are so deadly poisonous that they advertise their danger to enemies by being vivid reds and blacks or yellow and blacks. These frogs are the source of supply of the poison kurari to the local Indian tribes for use on the arrow tips in their blowpipes.

The AXOLOTL of Mexican freshwater pools is the Peter Pan of the amphibian world. It never grows up. The aquatic larval forms retain their gills and yet reach sexual maturity. They then breed without metamorphosing into adult salamanders. This is due to a lack of iodine in the thyroid gland which triggers off the life cycle changes.

Below: On land, the strong hind legs of a frog are used for jumping. The impact on landing is taken by the short front limbs.

Above: A Costa Rican flying frog caught 'in flight'.

Below: A painted reed frog utters a mating call by inflating its throat pouch.

Fishes were the first animals with backbones. Most fishes live in Sun-warmed surface waters, but some have adapted to icy polar waters and others exist in ocean trenches some 11 kilometres deep.

Fishes

Fishes evolved from invertebrate ancestors over 440 million years ago. They were the first group to have a backbone giving support to the body. The various fins projecting from the body are supported by parts of the skeleton. Fishes are aquatic cold-blooded vertebrates, virtually all of which breathe by gills. In the vast oceans and waters which cover about 75 per cent of our world, there are over 20,000 species of fish. They exploit every available habitat and show a great variety in size, shape, colour and behaviour.

The majority of fishes are bony fish of the class Osteichthyes, but there are some 600 species of cartilaginous fish of the class Chondrichthyes, which includes the SHARKS and RAYS. The most primitive fishes are the 50 or so species of jawless LAMPREYS and HAGFISHES of the class Agnatha.

The bodies of most fishes are covered with scales, which vary from the teeth-like scales of sharks to the plate-like scales of herrings and salmon. In the bony fishes the scales can be used to find out the age of the fish. Under a microscope a scale is seen to be made up of rings, each ring representing a year's growth. This is similar to the growth rings in tree trunks. In other fishes, such as the SEAHORSES and armoured CATFISHES, nearly all the body is covered in a protective bony armour. Some fishes have protective spines instead of scales.

Most fishes have two kinds of fins, paired and unpaired. The paired fins are the pectorals and the pelvics, one of each is on either side of the body. The unpaired fins are the dorsals on the back and the ventrals and anals on the underside. The tail (or caudal) fin in most fishes is the main means of propulsion.

How fishes swim

There is a great variation in the shape, size and use of the fins depending on where the fish lives and how it moves. The pectoral fins are usually

Above: Butterfly fish are colourful inhabitants of tropical coral reefs.
Right: There are 3 major classes of fishes (bony, cartilaginous, and jawless) that are recognised today.

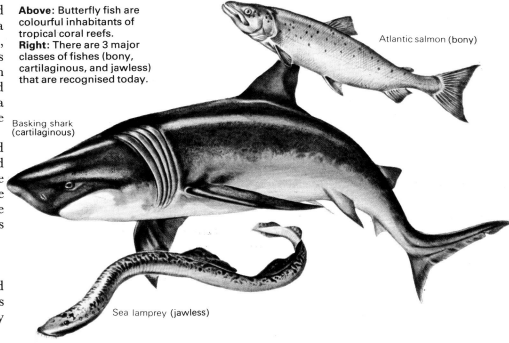

Atlantic salmon (bony)

Basking shark (cartilaginous)

Sea lamprey (jawless)

Reference

A **Angel fishes** are small fishes that have laterally compressed bodies. They are among the most beautiful fishes in the world and inhabit both freshwater (CICHLIDS) and tropical coral reefs. They show a fantastic range of colours and patterns in the 150 species. Some are up to 60 cm long. The related BUTTERFLY FISHES are usually much smaller.
Angler fishes have a moveable 'fishing rod' fixed to

their upper jaw that acts like a lure. The 'rod' is a modified spine of the dorsal fin. When a small fish is attracted to the lure, the angler gulps it into its mouth at a terrific speed.
Archer fishes can shoot a powerful jet of water at an unsuspecting insect lying up to a metre away. The victim drops into the water and is gulped down. They live in south-east Asia and Australasia.

B **Barracudas** are streamlined, fast-moving, carnivorous fish. Divers some-

times fear an attack by these 2-metre-long giants more than sharks.

Veil-tail angelfish

Butterfly fishes are colourful, tropical reef fishes closely related to ANGEL FISHES. Their small mouths and extended snouts are adapted for picking up invertebrates from cracks and crevices in coral. There are also unrelated freshwater species.

C **Carp** are large relatives of the GOLDFISH found in most temperate freshwater habitats. They can weigh over 25 kg. They have 4 barbels, 2 at each side of the mouth. Golden carp, bred by the Japanese, are spectacular fish for garden pools.

Catfishes are usually found in African, Asian and South American waters, although others live farther north. Most species have barbels to help find their food. They are either scale-less or are covered with bony protective plates (armoured catfishes).

Catfish

used for steering, or for slow movements as in seahorses. In the rays, the pectorals are huge and are practically the only means of locomotion. The fins are slowly flapped up and down and waves travel down the huge muscular fins to propel the animal along. The pelvic fins assist in keeping a fish steady. In the so-called FLYING FISHES, the pectorals are very long and almost reach the tail. When the fish has got up enough speed to shoot out of the water, these fins are expanded and the fish glides above the waves, the fins acting merely as aerofoils.

The majority of fishes that swim quickly use body movements and the tail fin. The body movements are caused by alternately contracting and relaxing body muscles known as myomeres. This causes the fish to wiggle from side to side in an S-shape. Water is pushed aside by the forward motion of the fish's head. It moves first to the left, then the right, the muscles contracting alternately down the fish's body to drive the fish forward. The sailfish is reckoned to be the fastest fish.

Below: To swim forwards, most fishes move their body from side to side by the alternate expansion and contraction of body muscles, the myomeres. The various fins aid movement also. The illustration of a shark swimming shows how the body is curved and how the flexure moves backwards from the head to the tail.

How fishes feed

Movement is important to fishes, firstly to escape being eaten by bigger fishes and other marine creatures, and also in finding food to eat. A large number of bony fishes are fish-eaters (piscivorous), such as the COD, PERCH, BASS and PIKE, and they usually have strong pointed teeth with which to seize their prey. The pike's large mouth simply bristles with teeth, not only in the jaws, but also on the roof of the mouth and the tongue. The pike lurks in a clump of vegetation and as soon as a victim, usually a fish, comes within reach it is snapped up. Hungry pike will even eat their own kind. They are the true cannibals of their world.

Another fish renowned for its ferocity is the PIRANHA of the rivers of South America. They live in groups and have sharp powerful jaws equipped with sharp cutting teeth. Although their usual diet is smaller fishes, they will attack any animal unlucky enough to fall in the water. It is then cut to pieces in a very short time.

Right: The sand tiger shark shows his rows of sharp, pointed teeth.

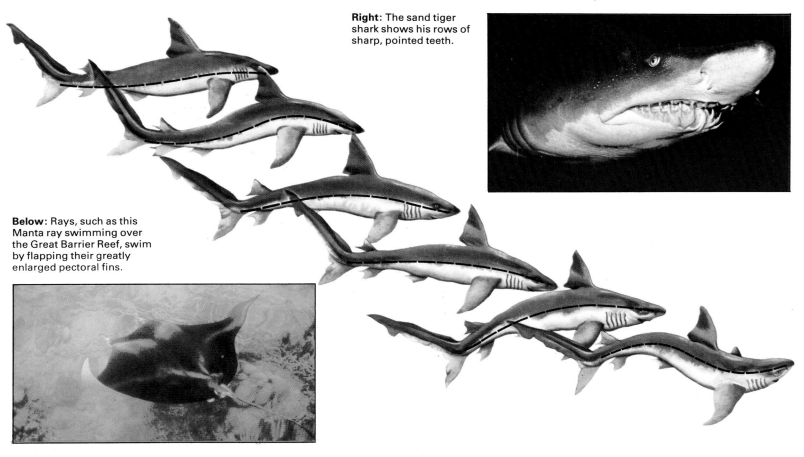

Below: Rays, such as this Manta ray swimming over the Great Barrier Reef, swim by flapping their greatly enlarged pectoral fins.

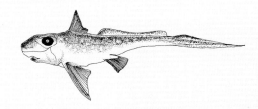

Chimaeras are strange-looking relatives of SHARKS and RAYS. The upper jaw is fixed to the skull and the male has a clasper in front of his eyes that probably has some courtship function. In some species, the long dorsal spine carries a sting.
Cichlids are an aggressive family of some 600 species found in tropical freshwaters of the world, but mainly in South America. Many species are very popular with aquarium owners, especially angel fish, the pompadour fish and the MOUTHBROODERS.

Chimaera monstrosa

Codfishes occur chiefly in the northern seas and are, with the haddock, pollack and whiting, some of the world's most valuable food fishes. Most species live near the sea bottom and so are trawled by fishermen.

The Atlantic cod is the largest of the 150 species in its family (Gadidae). It has a small whiskery barbel under the chin.
Coelacanths are living fossil fishes. They were thought to have become ex-

tinct some 65 million years ago until 1 was trawled up off East London, South Africa in 1938. The leg-like fins can move in all directions and are probably used to stir up mud on the sea floor when searching for prey.

D **Dogfishes** are small SHARKS. In Europe they are fished commercially. The common dogfish is an inshore, shallow-water species.

E **Eels,** of which there are over 300 species, are mainly marine and live in shallow water. The majority are scale-less and the dorsal and anal fins are continuous with the tail fin. Most eels

Eels

Sharks are the other notable flesh-eaters and these cartilaginous fishes have several rows of teeth in their jaws. As they are worn down at the front of the jaw they are replaced from behind. The teeth of the great white shark, or man-eater, are the most formidable and powerful of all sharks. The white shark feeds mainly on fishes, but porpoises, water birds and even sealions are taken, as well as humans.

Many fishes feed on small aquatic insects and their larvae. TROUT will leap out of the water to catch flies and gnats near the surface. The ARCHER FISH are so-called because of the fact they 'shoot down' insects with jets of water.

Fishes breathe through gills and these are always found between the mouth and the beginning of the food tract or gut. This region is called the pharynx. In sharks, the walls of the pharynx have several narrow gill slits which open to the outside. Water is taken in through the mouth and passed over the gills and out again. In bony fishes, the gill slits have a moveable cover

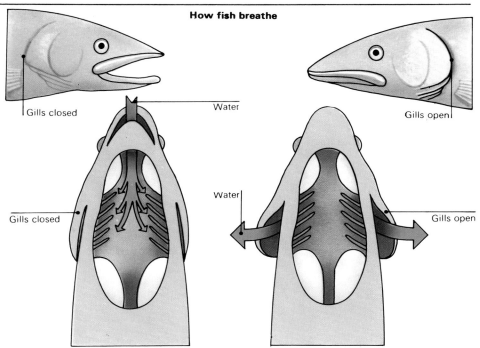

How fish breathe

Gills closed · Water · Gills open · Gills closed · Water · Gills open

Below: The archer fish of south-east Asia gets its name from its curious feeding method. On seeing a fly or insect over its watery home, the fish takes aim and squirts a drop or two of water from its mouth at its victim. An adult fish can eject a jet of water over a distance of 90 cm.

Above: When the mouth of a fish opens, oxygenated water is drawn in. The fish extracts the oxygen and the water is then passed out through the gills.

Below: The piranha of South American rivers swim together in schools. They can grow up to 38 cm long. Their jaws are armed with sharp cutting teeth that can

quickly slice up the flesh of other fish or mammals.

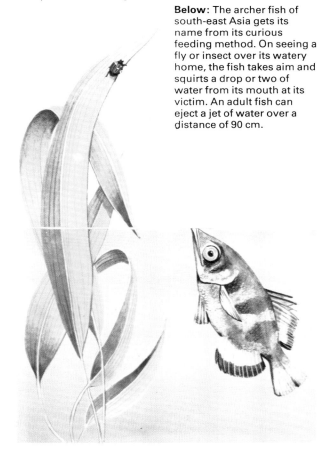

have a transparent, ribbon-like larva known as a leptocephalus. It eventually changes into a small eel.

Electric eels are freshwater South American fishes of the family Electrophoridae, not closely related to true EELS. They have electric organs which enable them to establish electrical fields around their bodies as an aid to locating enemies and stunning prey. The anal fin helps the fish to move.

F Fighting fishes from Thailand are bred and used for sport. The wild ancestor is small and dull compared to aquarium breeds. Varieties in captivity range from cream-coloured with red fins to a purplish-blue. Only the males are aggressive, especially during the breeding season. The male builds a bubble nest for the eggs, and then he guards the nest and young.

Flatfishes are several species that change (undergo skeletal metamorphosis) from being normal larval fish with eyes on each side of the head to a flatfish that lives on the bottom of the

Plaice

sea and spends the rest of its adult life there. The eye on the underside migrates to the opposite side of the head and the mouth becomes twisted. The blind side does not develop pigment but the upper side is able to vary its colour pattern with the surroundings. Many flatfishes are valuable food fishes, such as the flounder, plaice, turbot, sole and halibut.

Flying fishes are small, slender, bony fishes that are found in the surface waters of the oceans. Large pectoral fins enable them to glide through the air. Some species cover over 150 metres in one flight.

G Goldfishes in the wild are plain, brownish CARP-like creatures. Today, many different types of domestic goldfish have been bred, mainly pioneered by the Japanese.

called the operculum. The gills are supported by a special part of the skeleton called the gill arches. As water is passed over the gills, oxygen passes through the thin walls of the gills into the bloodstream, and the waste carbon dioxide passes from the blood through the gills and into the water.

Fishes that breathe air

Most fishes when taken out of water die because their gills cannot function properly. Although air contains more oxygen than water, the gills cannot extract it, and so the fish suffocate. There are fishes, however, that can breathe by other methods. EELS often make considerable journeys over wet ground. This is because they are able to breathe through their skin as long as it is kept moist, just like frogs and newts.

Climbing perch, FIGHTING FISH and certain catfishes of Africa and Asia have accessory breathing organs off the gill chambers. The LUNGFISHES, as their name suggests, have lungs and they can breathe air. The lung is an outgrowth of the gullet and has evolved from the swim bladder, whose function in most fishes is to give buoyancy. The ability to breathe air allows lungfishes to survive periods of drought in Africa, Australia and South America.

Above: An African lungfish can lie dormant for weeks or months in its mud cocoon for the duration of the dry season. It has a tiny breathing hole to the surface.
Below left: After spending years at sea maturing and feeding, salmon navigate their way back to their freshwater birthplace possibly by the Sun and a sense of smell.
Below right: A cleaner fish here provides a 'doctoring' service to a coral cod by eating bacteria, fish lice and fungus from its skin and inside its mouth.

Behaviour and breeding in fishes

In adapting to various habitats some strange aquatic partnerships occur, for instance between Damselfishes and tropical sea anemones in the Pacific Ocean. Even more extraordinary are the barber (or cleaner) fish who clean external parasites from the skins and gills of other fish. These fishes are usually brightly coloured as if to advertise their presence to 'customers'. They certainly help to maintain the well-being of other fishes and are especially abundant in tropical regions. Some fishes, such as FLATFISHES, can change colour to match their backgrounds.

Fish migration is associated with the reproductive cycle of the species and with the availability of food. Commercially exploited fishes, such as the blue-fin tuna (tunny) and the Atlantic SALMON have been well studied. The migration of salmon is complex, the fish leaving their birthplace in the headstreams of freshwater rivers to spend several years at sea feeding and maturing. They then return to the river of their birth to spawn. Experiments have shown they have an excellent sense of direction and their acute sense of smell enables them to detect their original birthplace. The European and American eels are famous for their migrations from freshwaters to the Sargasso Sea, the young then

Goldfish

H **Hagfishes** are jawless fish that remain buried in mud, gravel or sand by day. At night they emerge to scavenge on dead animals or any waste organic material. Some are parasites on living fishes. They have lost their eyes and hunt for food by touch and smell.

Herrings are silvery fish that travel in huge shoals and are very important food fishes. They make unpredictable seasonal migrations and move into shallow waters for spawning.

L **Lampreys** attach themselves by suckers to the sides of fishes. They rasp through the skin and flesh and drain the victim of blood. The adults usually die after they have reproduced.
Lantern fishes are small, deep-sea fishes that have light-producing organs, or photophores, along their sides. They migrate daily from 1,000 metres or more to feed near the surface.
Lungfishes are air breathing relics from the DEVONIAN PERIOD (*see page 4*). African and South American lungfishes have eel-shaped bodies. They have 2 lungs and aestivate (hibernate) in dried mud with a breathing hole through periods of drought. The Australian species has only a single lung and dies quickly if its surroundings dry up.

M **Man-eating shark** is the term given to those SHARKS that are known to attack man. It is most often used to mean the great white SHARK, but others include the mako, tiger, sand, and hammerhead.
Mouthbrooders are species of CICHLIDS that brood their eggs inside the mouth. It is usually the male that does this. Even after hatching the young are held in the mouth for a short time. The young fish are thus guarded and whenever they stray too far or danger threatens, they either swim into, or are sucked up by, the parent's mouth.

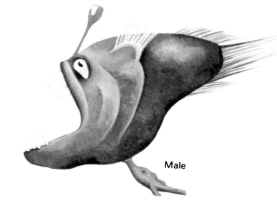

Left: A male seahorse broods the fertilized eggs in a pouch on his belly, and gives 'birth' 5 weeks later by contracting his body in jerks.
Right: A female, deep-sea angler fish with a parasitic male attached to her underside. His only function is to fertilize her eggs. This arrangement avoids the need to locate a partner in the dark sea depths.

Male

Right: A mouthbrooding cichlid protects her eggs by carrying them in her mouth. Even after hatching, the young do not leave the shelter of her mouth until they are able to look after themselves.

Below: Niko Tinbergen, the German naturalist and ethnologist, made the first studies of the courtship of the 3-spined stickleback. Sign stimuli were shown to be the most important. A male will attack another male if it has a red throat and encroaches on his territory.

drifting with the ocean currents back to the mouths of rivers.

In the breeding season many bony fishes, such as HERRINGS, haddock and cod, gather together in shoals. The females lay their eggs and the males shed milt or sperm onto the eggs and fertilize them. Millions of eggs die, with perhaps only one of those fertilized eggs developing into a mature adult fish. A female turbot, for example, produces about nine million eggs each breeding season.

Sticklebacks and fighting fish show more parental care and build nests for their eggs. Some fishes, such as the MOUTHBROODERS, carry their eggs inside the mouth. In the American CATFISH it is the male who does this. Male pipefishes and seahorses carry the fertilized eggs inside a special pouch on their belly. When the eggs have hatched, the male looks as if he is giving birth as he bends and stretches to eject the miniature seahorses from his pouch.

Female sharks and rays either give birth to live

Red-bellied, blue-eyed male courts a female.

He leads her to his nest which he has built.

He nudges her gently to get her to lay her eggs.

She is now driven away from his nest.

He wriggles into the nest to fertilize the eggs.

The male guards and fans fresh water over the eggs.

P Pike are voracious carnivores, usually at the top of most freshwater food chains. The northern pike occurs in North America and Eurasia where it can grow to over a metre long and weigh about 20 kg.
Piranhas are small, South American freshwater fishes that immediately bring the word 'killer' to mind. Probably only 4 of the 20 or so species deserve this title. They mainly catch small fish, but if larger animals, such as a rodent like the capybara, get into difficulties in water, then hundreds will gather and quickly reduce the animal to a skeleton.
Plaice, see FLATFISHES.
Pufferfish have tiny mouths armed with heavy teeth and short, deep bodies that are often covered with spines. Most live over coral reefs. When pulled from the water they swallow air, or if threatened they will take in water. Thus they become balloon-shaped and are difficult for any predator to swallow. The porcupine puffer fish is covered with spines so that this forms an even more impenetrable barrier.

Porcupine pufferfish

R Rays are flat-bodied cartilaginous fishes with large pectoral fins that are flapped in a wing-like fashion to move through the water. The stingrays and whiprays have venomous spines. The manta, or devil rays, are docile but huge, measuring up to 7 metres across the wing tips. The giant mantas are plankton feeders and do not attack man.
Remoras are fishes that have a sucker disk on the top of the head which has evolved from the dorsal fin. A remora uses this sucker to attach itself to fishes such as sharks, turtles and whales. It does not harm the host and takes parasitic crustaceans from the host's skin.

S Salmon are popular fish with anglers because they put up a spectacular fight. They are hatched in

Below right: The stonefish have venomous glands at the base of their spines. They lie perfectly camouflaged on the sea bed, and if a bather steps on one, poison is injected into the wound by the pressure of the foot on the bag-like glands. Effects vary from intense pain to cases of death occurring in 6 hours.

Right: The lateral lines on this tench are sense organs that run along the sides of its body. Along these lines, sensitive cells called neuromasts pick up pressure waves and translate them (via the brain) to indicate the direction and the size of the object producing the waves.

young, or lay a small number of eggs. These develop inside a horny, protective leather case.

The senses of fishes

The sense organs of fishes are not unlike those of higher vertebrates, but the adaptation to a water habitat means some are greatly accentuated while others are less developed. Most fishes have nostrils but these are never used for breathing purposes. They open into little sacs into which the water enters and can be smelt. Many fishes, such as sharks and piranha, hunt by smell, and salmon use smell to find their birthplaces.

Taste is another sense and it is closely linked to that of smell. Fishes have taste buds not only on their tongue but in many species, such as CARP and STURGEON, they are found on the head and body. CATFISH have tastebuds on the barbels that surround the mouth.

The eyes of fishes are adapted for underwater vision and are usually placed on the sides of the head. Very few fishes have binocular vision. The hammerheaded shark has its eyes at the ends of its 'hammer', and this probably improves its vision. One curious fish from Central America has eyes that stick out. Each eye is divided into two halves, the upper half is focused for seeing in air and the lower half for underwater.

Fishes hear vibrations in the water. The sound waves are sometimes magnified by the air bladder and carried to their inner ear. Fishes also have a sense organ which is not found in higher vertebrates. This is the lateral line system. It consists of a line of tiny pores along each side of the head and body. Nerves run to the brain from each pore and it seems that this system is nature's version of radar. When a fish swims near a rock, the vibrations sent out by the moving fish hit the rock, are reflected, and are picked up by the lateral line pores. This explains why blind cave fishes can move about in total darkness.

---

freshwater headstreams, migrate to sea and return several years later to spawn and then usually die.
Scorpion fish are also called dragonfish, turkey fish or lion fish. The name comes from the venomous nature of the long spines on their fins. The related stonefish is the most poisonous fish known.

Scorpion fish

Seahorses are strange fishes that move through the water with the body vertical. They have a bony armour and their horse-shaped head has a tubular snout. The female lays the eggs and the male broods them.
Sharks are some 200 species of cigar-shaped fishes, many of whom are predatory and some are man-eaters. Whale sharks and basking sharks are the largest living fishes, reaching almost 20 metres; they feed on plankton.
Sticklebacks are small, aggressive fishes with armoured bony plates down the sides of the body. They also have spines on the dorsal ridge. In the breeding season the male develops a bright red belly and constructs a nest. He attracts a female by a zig-zag dance. The female lays her eggs in his nest and the male immediately fertilizes them and then chases her away. He brings up the young fishes alone.

Stonefish, see SCORPION FISH.
Sturgeons are fishes of northern temperate waters that can grow to almost 1,000 kg. They have 5 rows of plate-like scales along the sides of their bodies. Sturgeon's eggs are made into caviar.

T **Trout** are closely related to SALMON and are usually restricted to fresh waters although some, such as the sea trout, migrate between sea and river.

Shovel-nosed sturgeon

Arthropods are the most numerous of all animals. They include insects — the largest animal class of all — crustaceans, arachnids, centipedes and millipedes. The number of recorded insects is increasing as new species are discovered.

Arthropods

Members of the arthropod (jointed limbs) group are the most numerous and widespread in numbers of species and individuals. They are the only invertebrate group to have successfully adapted to living on land and include the only other animals, apart from the vertebrate birds and bats, to have become modified for flight. More than 800,000 species have been described for this group and this is about 80 per cent of all known animal species.

All these animals, whether LOBSTER, CENTIPEDE or BEETLE have a hard shell that forms an outer skeleton (the exoskeleton). The shell is rather like a suit of armour in that it is jointed and the muscles are arranged across the joints so that the parts of the exoskeleton are able to move in relation to each other. The shell is almost impermeable to water in both directions. This is very important to those arthropods on land, especially the large group of INSECTS, as it prevents them from drying out.

One disadvantage of the shell is that it restricts growth, and the developing arthropod is always outgrowing it. Every so often the shell is shed or moulted. Usually the animal splits its old cuticle (skin) and then pulls itself out. A new cuticle has already grown and while it is relatively soft and pliable, the animal stretches and increases in size. Shortly afterwards, the cuticle hardens and again becomes an armour-like exoskeleton. No further increase in size takes place until the next moult. When an arthropod has just moulted it is in great danger as it is defenceless with a 'naked' body, and also it cannot move very far.

Movement of the millions

Some arthropods such as BEES, dragonflies and lacewings, all of which are insects, are great fliers. Others like spiders and silverfish move around on the ground on their jointed legs. There are also thousands that live and move about in

Crustacea: Lobster

Arachnida: Scorpion

Insecta: Grasshopper

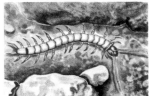
Merostomata: King crab

Symphyla: Symphylan

Thysanura: Silverfish

Onychophora: Velvet worm

Collembola: Springtail

Diplopoda: Millipede

Pauropoda: Pauropod

Diplura: Bristletail

Chilopoda: Centipede

Protura: Proturan

Above: The arthropods form the largest phylum in the animal kingdom, containing more than 800,000 species. All the animals in the 13 classes have their bodies divided into segments and have an outer skeleton called the cuticle.

Above: A beautiful swallowtail butterfly. It gets its name from the tail-like extensions of the hind wings.

Right: A vertical take-off by a green lacewing showing its controlled manoeuvres and delicately veined wing membranes.

water, such as crabs, lobsters, water beetles and copepods, and their legs have adapted for this life.

In arthropods, the number of legs varies and some have become modified for other functions. In a few cases the number is as few as two pairs, as in a preying mantis, but most insects use three pairs of legs and spiders use four pairs. CRABS, although they have five pairs of legs, use only three in crawling sideways, the other pairs being claws adapted for feeding and grasping, and for swimming. Water beetles have flattened, paddle-shaped legs for swimming, while the third pair of legs in FLEAS and GRASSHOPPERS have become specialized for jumping.

Insects first flew some 200 million years before reptiles and birds took to the air. Over 320 million years ago in the Carboniferous period there were immense dragonfly-like creatures with a wingspan of up to 76 centimetres.

Insect wings are attached to the THORAX via couplings that act like a series of ball and socket joints. This enables the animal to fly sideways, backwards, upside down, swoop, climb vertically

Right: Giant millipedes do not have millions of legs. They can have up to 355 pairs, but usually less than 100 pairs are present, depending on the species. The longest species can measure up to 28 cm in length and 2 cm in diameter.

They range in size from the microscopic fungus beetle to the gigantic tropical atlas and hercules beetle.
Bristletails are small- to medium-sized wingless insects with 3 bristles at the tip of the abdomen. They have chewing mouthparts for feeding on decaying plant material.
Bugs belong to the order Hemiptera which means 'half wing'; only half of the front pair of wings is thickened. They have piercing and sucking mouthparts and can do great damage to plants on which they feed.

White admiral butterfly

Butterflies, together with MOTHS, form the large insect order Lepidoptera, which contains more than 120,000 species. Many beautiful and large specimens are collected and have now become rare. Butterflies have clubbed antennae and fly during sunny periods of the day. They undergo complete metamorphosis – from egg, larva or caterpillar, pupa or chrysalis, to adult.

C **Centipedes,** of the class Chilopoda, have flattened bodies with a pair of legs on each segment, except for the last two. There are some 5,000 species in the world.
Chilopoda, see CENTIPEDES.
Cicadas are more often heard than seen. These 1,500 species of insect sing to each other by vibrating a small drum-like membrane. They are the largest members of the order Homoptera.
Cockroaches are very ancient insects, with most of the 3,500 living species coming from the tropics. They have flat bodies and long, slender running legs. They hide in cracks by day

Male cockroach

and come out to feed at night.
Crabs, of the class Crustacea, are relatives of the

and hover. Power is applied in different ways depending on the species but usually comes from two pairs of large muscles.

How arthropods feed

Within the arthropod grouping almost every conceivable material is used as a food source by one or more insect groups. There are fish-feeders, plankton-feeders, wood-feeders, blood-feeders, dung-feeders and sap-feeders to name but a few.

Some limbs have been modified to act as feeding organs. These lie on or close to the head. The pincers of crabs and scorpions catch the food, while other limbs break it up and pass it to the mouth. Beetles, ants, wasps, grasshoppers, locusts, earwigs and dragonflies on the whole have biting mouthparts, with jaws or mandibles to cut and crush the food. All true BUGS and flies, as well as most butterflies and moths, have sucking mouths but the actual structure of the mouthparts varies. Houseflies are only able to suck up liquid food. BUTTERFLIES and moths that feed on the nectar of flowers have a tubular mouth (proboscis) which is coiled up under the head when not in use. It unrolls and acts rather like a drinking straw when the insect is feeding. Bugs, such as greenfly, have sharp piercing mouthparts to suck up the sap from a plant's tissues. MOSQUITOES and fleas pierce their victim, usually a mammal or bird with their needle-shaped jaws and then suck up the blood.

SPIDERS are mainly carnivorous. They usually capture victims, usually small insects or other arthropods, by a web or trap. Enzymes, secreted by the spider, are injected into the prey and its tissues are broken down so that it can then be sucked up by the spider's mouth. Some spiders possess fangs which inject poison into their prey to paralyse or kill it.

Group behaviour

The behaviour of arthropods is extremely varied but basically it is designed to ensure the species survives in order to reproduce. Most behaviour is instinctive, the behaviour patterns being passed on from generation to generation. A limited amount of learning is possible in some insects. Cockroaches can be trained to go right or left by punishing them when a wrong turn is made.

In the insects we find the most highly-

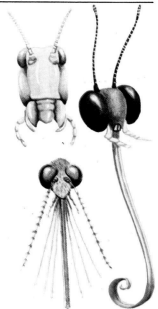

Above: A grasshopper's mouthparts *(top)* are designed for cutting. A butterfly's tube *(right)* sucks up nectar, and a mosquito *(bottom)* has a bloodsucking tube.

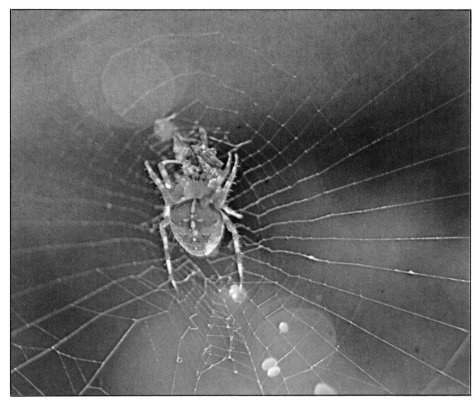

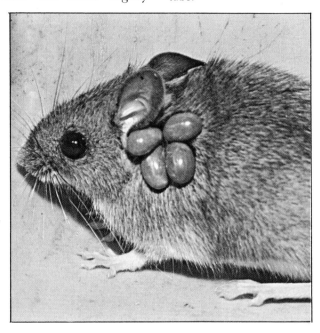

Left: A female garden spider is seen here eating her mate. In the animal world, the female of the species is often deadlier than the male.

Above: The ticks on this little wood mouse have swollen up by feeding on their host's blood. They leave their host only to breed.

LOBSTER. They usually run sideways or 'crab-wise', although they are able to move in any direction. Although mainly seashore or marine in habitat, a few live in freshwater. Their pincers are used for feeding, defence and courtship.
Crayfish are freshwater lobsters found mainly in temperate streams and rivers.
Crustaceans include SHRIMPS, prawns, LOBSTERS, crayfish, CRABS, BARNACLES and krill (the food of baleen whales). The 26,000 species are mainly marine or freshwater.

Krill

D **Daddy-long-legs,** see HARVESTMAN SPIDER.
Diplopoda, see MILLIPEDES.

F **Fiddler crabs** live mainly in the tropics. The males have one absurdly big claw, termed the fiddle, which is used for defence and courtship. Females have no fiddle.
Fleas are small wingless INSECTS that feed on the blood of warm-blooded mammals and birds. They have a compressed body covered with spines that point backwards, thus allow-ing them to move freely between an animal's feathers or hair. An infested animal may have many thousands of fleas on its body at any one time. The majority of the 1,800 species live in the tropics.
Flies are INSECTS with only one pair of wings. They are very numerous and widespread. They show complete metamorphosis, the larvae often being called maggots.

G **Grasshoppers,** totalling over 10,000 species, are either short- or long-horned. Each species has its own distinctive sound. Shorthorns make their sound by rubbing the femur of their hind leg against the hardened area on the front wing. Longhorns rub their front wings together.
Greenfly, see APHIDS.

H **Harvestman spiders** are seen most often at harvest time, and are often termed daddy-long-legs.
Horseshoe crabs, or king crabs, are neither crabs nor CRUSTACEANS. They are ARACHNIDS, the group to which spiders and scorpions belong. Their bodies are pro-

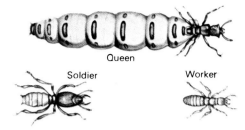

Queen

Soldier   Worker

Left: Termite society is divided into workers, soldiers and the colony's king and queen.
Below: Air in the living quarters of the termite mud nest is heated by the activity of millions of workers. This air rises to the top of the nest. Here fresh, cooler air enters from the outside and circulates down the nest.

Fresh air

Stale air

organized social behaviour. Although the majority of insect species lead solitary lives, two groups are social. These are the TERMITES, or white ants, of the order Isoptera and the BEES, wasps and ANTS of the order Hymenoptera. Here parents and offspring live together in a community in one nest and actually co-operate in running the home. An ant colony is headed by a queen, with various castes (groups), such as soldiers or workers, performing different tasks.

Many moths and butterflies migrate thousands of kilometres annually to escape harsh climatic conditions and food shortages. The most famous and well-tracked insect migration is that of the monarch butterfly of America. LOCUSTS build up to plague numbers and move off as a swarm in a certain direction, devastating all the vegetation in their path, but do not return. This one-way movement is called emigration.

Life cycles and reproducing

In most arthropod species there is a different set of instinctive behaviour patterns for the two sexes. The female's role is to deposit and protect the eggs. Courtship displays, where the sexes come together for mating, are brief. A male fiddler crab waves his huge claw to attract a female. Many jumping spiders wave their legs and dance from side to side to attract a mate. A small male preying mantis when attempting to mate with his chosen larger female must approach with great caution or he will get his head bitten off by the predatory female.

Like most invertebrates, arthropods lay large numbers of small eggs. Usually the young hatch at an immature stage and must feed to grow and develop further. The larvae of aquatic species are free-swimming and undergo a gradual change into an adult. Caterpillar larvae are so different from the adult butterfly that it is not possible to have a gradual change. A larva surrounds itself

tected by an armour-plated shell and they have survived almost unchanged for 500 million years.

Insects account for about 80 per cent of all the known kinds of animals. There exist almost 1 million species. Some beetles and midges are the size of a pin head while stick insects can measure about 30 cm, and some moths have a wing-span this size. The presence of wings is the best way of distinguishing insects from other arthropods, although lice and FLEAS have lost

Hercules beetle

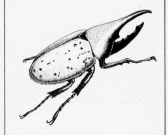

theirs. The head has one pair of antennae and 3 pairs of mouthparts. The 3-segmented thorax has a pair of legs on each segment.

The abdomen has 11 segments.

Ladybirds are useful BEETLES which feed on pests, such as APHIDS. Most of the 5,000 species are yellow, orange or red with black spots.
Lobsters are large CRUSTA-CEANS, related to shrimps. Young lobsters look like small shrimps. Later they develop huge pincer claws, settle on the sea bed and become scavengers.
Locusts are destructive short-horned GRASSHOPPERS. Each continent has one or

more species of locust that at times builds up an enormous population that then migrates in incredible swarms. The reasons why are still not fully understood.

Ladybird

Merostomata. This class contains the HORSESHOE CRABS.
Millipedes, of the class Diplopoda are ARTHROPODS with a horny outer layer that forms a hard armour that is used for burrowing. They usually have more legs than CENTIPEDES, although the longest has no more than 200 legs. The 8,000 species worldwide are usually vegetarian.
Mites are related to TICKS, and are quite tiny. Most species live in leaf litter. Segmentation typical of ARTHROPODS is reduced or absent. They have four pairs

with a protective material and transforms itself into a pupa. This is referred to as its 'quiescent' stage, and indeed it is externally. After a time a sexually mature adult emerges. These changes are termed metamorphosis, which means change in form. In the many insects that go through this process, there is a marked division of labour among the different phases of the life history. A caterpillar hatches complete with chewing jaws and immediately starts to feed on leaves. Its role is feeding and in a short time it has eaten large quantities of food and grown quite rapidly. The pupal resting stage is the transformation stage. The winged adult butterfly that emerges has no biting or chewing jaws but sucks nectar with its coiled tubular mouth. Its chief adult role is to distribute the species by flying to new and suitable habitats and finding a mate. Some adults do not even feed, mating soon after emerging from a pupa and dying when the eggs have been laid.

Some primitive insects, such as silverfish, hatch out as small-scale replicas of their parents. They grow to adult size by feeding and moulting. In COCKROACHES and LOCUSTS, a nymph hatches out of the egg. It looks like a miniature adult, except that its wings are undeveloped. As it grows and moults the wings gradually become more identified until the fully mature stage is attained. This sequence of egg-nymph-adult is known as incomplete metamorphosis.

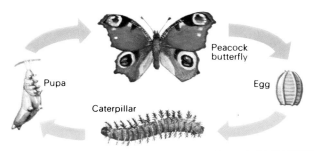

Peacock butterfly
Pupa
Egg
Caterpillar

Left: The peacock butterfly's life cycle is an example of complete metamorphosis. This is where the egg hatches into a form entirely different from the adult. This larva, or caterpillar, passes through a pupal or chrysalis stage where it changes into the adult form.

Above: Malayan fiddler crabs defend their small territories of sand by signalling with their large claw. The claw is also used in courtship displays.

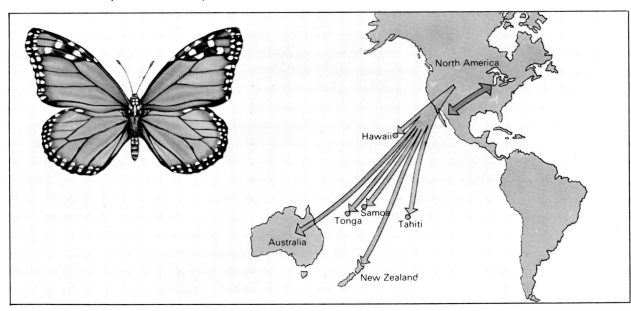

North America
Hawaii
Samoa
Tonga
Tahiti
Australia
New Zealand

Left: The North American monarch butterfly is one of the few species of insects that make annual 2-way journeys comparable to the distances covered by some birds. Some travel over 3,000 km between their summer breeding places and winter feeding ranges in the south. In the last 130 years the butterfly has expanded its range westwards across the Pacific, probably using ships for rides.

---

of legs.
Mosquitoes inflict more direct harm to man than any other known INSECT. The 2,000 species are distributed worldwide. A mosquito female needs a meal of blood after mating and before she lays her eggs. When they bite man, diseases such as malaria and yellow fever can be transmitted. Control methods, such as chemical spraying or draining the breeding sites, have had some success.
Moths, together with BUT-TERFLIES, belong to the insect order Lepidoptera. Most moths are nocturnal. The males have hairy bodies and feathery antennae. At rest the wings are held horizontally and not vertically as in butterflies.

Mosquito

P **Pauropoda.** This class contains tiny ARTHROPODS about 1 mm long that live mainly in the soil.
Peripatus is a velvet worm of the class Onychophora. It lives under logs and stones in the tropics and is up to 15 mm long.
Proturans are the class that contain tiny 12-segmented animals about 1 mm long that are found in leaf litter. They use their fore-legs as feelers.

S **Scorpions** are related to spiders and are most abundant in the warmer reg-ions of the world. Some measure about 15 cm in length. The tip of the abdo-menal tail has a powerful poison gland which stings or paralyses enemies or prey. Prey are then sucked dry of their juices.

Hawk moth

Above: The prominent compound eyes of a bluebottle are seen here. Each eye is made up of about 4,000 6-sided facets.

Below: The feathery feelers of this male Thailand Atlas moth can trace the scent of a female up to a kilometre or more away.

Left: A cross-section of an insect's compound eye. Each facet consists of a tiny lens, a light-transmitting system and sensitive retinal cells. Each facet registers a fragment of the total picture seen so that the whole eye builds up a jigsaw mozaic of the scene. It is probably a blurred image because insects cannot focus.

Senses of arthropods

The most important sense of a species depends on the kind of life it leads. CRUSTACEANS and insects are arthropods that have a pair of compound eyes. Most of the other arthropods have only simple eyes. Actually, most insects possess both simple and compound eyes. Simple eyes have no focusing mechanism but probably only measure light intensity. A compound eye is composed of hundreds or even thousands of light-condensing units. It does not give as good an image as the human eye. Insects probably see images similar to a human being looking at a newspaper photograph through a magnifying glass. The compound eye can detect the slightest movement of an enemy or prey.

Smell bring the sexes together, finds suitable egg-laying sites or food, and often enables a species to identify its own kind. Male MOTHS use their feathery sensory antennae to home in on female moths that are giving off chemical scents. They can locate a female up to three kilometres away. Social insects, such as ants or bees, can not only detect members of their own species but can sniff out and attack an intruder from a rival colony.

Taste receptors occur mainly in the mouth and mouthparts, at the tip of the antennae and on the lower parts of the legs. When a fly lands on your jam sandwich he is actually tasting the food with his feet!

Hearing is unequally developed in the arthropod group. Many species are probably deaf, although the insects are the best equipped at hearing. Their range covers a wider frequency band than that to which the human ear is sensitive. The 'ears' of an insect are either sensitive hairs, or tympanal organs situated on the first abdominal segment. When a sound hits this tympanal organ, an air sac beneath it transmits the vibrations to the brain.

Sea spiders are a small marine group of the class Pycnogonida that have long narrow bodies and 4 to 7 pairs of legs. They are not true spiders.

Shrimps are CRUSTACEANS, closely related to prawns and lobsters. They are active swimmers finding food mainly by scavenging.

Spider crabs are peculiar CRUSTACEANS with long legs. Their body is often covered with warts or spines. The giant spider crab from Japan may measure more than 3 metres across the claws, and is the largest crustacean.

Spiders belong to the class Arachnida and are widespread, ranging in size from less than 1 mm to 25 mm long. They have chelicerae that bear poison fangs, and have silk glands in their abdomen. Four pairs of long legs are attached to the abdomen. Many species have excellent sight.

Termites are often called 'white ants' although they are not ANTS at all. Ants have a waist between the thorax and abdomen whereas the join is broader in termites. These social insects build huge colony mounds up to 10 metres high.

Thorax is the second of the three sections of an INSECT. It

Leaf-curling spider

bears the legs and wings.

Ticks are closely related to MITES and belong to the Arachnida class. They are all parasitic at some stage in their lives. They gorge themselves on blood and often transmit diseases.

Wasps belong to the same order of INSECTS, the Hymenoptera, as ANTS and bees. Many species are social but some are solitary, such as the potter wasps.

Wood lice are land CRUSTACEANS, their bodies being protected by hard plates. When certain species are disturbed they are able to roll into a ball, like tiny armadillos. They prefer damp places and come out to feed at night.

The invertebrate world includes many other species besides the arthropods. They range from the simplest animals—microscopic protozoans, which some botanists claim are plants – to the giant squid and the Portuguese man-o'-war.

Other Invertebrates

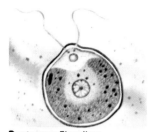

Protozoa: Flagellate

Porifera: Bath sponge

Coelenterata:
Portuguese man-o'-war

Platyhelminthes: Liverfluke

Aschelminthes: Roundworm

Mollusca: Edible snail

Annelida: Earthworm

Echinodormata: Sun starfish

Hemichordata: Acorn worm

Tunicata: Sea squirt

Nemertina: Ribbon worm

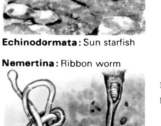

Brachiopoda: Lamp shell

The previous chapter mainly dealt with the jointed-legged arthropods, but the invertebrate world exhibits an enormous variety of animal life. Other groups include the PROTOZOANS, microscopic SPONGES, JELLYFISH, FLATWORMS, ringworms, snails, worms and STARFISH.

The simple, single-celled *protozoa* play an important part alongside microscopic plants in aquatic habitats, by providing the basis for all food chains in the sea. The many-celled sponges evolved from colonies of single-celled protozoans, but the former is not a very successful group, mainly because they have no nervous system linking the activities of various parts of the body. The COELENTERATES and flatworms are more advanced because they have evolved a simple nervous system. These groups also show the beginnings of a digestive system. The segmented ANNELIDS (worms) are adapted for swimming and burrowing. They have a better developed nervous and blood system.

Above: The numerous species in the various invertebrate phyla illustrated above have adapted to fill various niches in water or on land. Some species, such as the lamp shells, are survivors from the past while others such as the molluscs are very numerous and successful all over the world. The tunicates and hemichordates link the invertebrates with the vertebrates.

A major invertebrate group is the MOLLUSCA, most of which have a heavy shell and are slow-moving, for example garden snails.

There is a strange assortment of creatures at the more complex end of the invertebrate scale. The ECHINODERMS are very distinctively symmetrical, including the familiar starfish, BRITTLE STARS, feather stars, sea urchins and SEA CUCUMBERS. They all have spiny skins and the five-rayed star pattern can be found in almost all the species. The primitive tunicates and HEMICHORDATES are closer to the vertebrates than the invertebrates because in the larval stage in their life they possess a rod of inflated cells (the notochord) that acts as a support for the body, thus showing some similarity to the spinal column in vertebrates.

Senses and movement

Sense organs and methods of movement in the numerous groups vary according to their habitat

Reference

A **Amoeba,** a single-celled PROTOZOAN, is often incorrectly thought of as being the simplest form of life living today. It lives mainly in water but is sometimes found as a parasite in man.

Annelids, or worms, have their muscular segmented bodies covered in a thin skin and bristles. There are over 6,800 species in the 3 classes (BRISTLEWORMS, EARTHWORMS, and LEECHES).

Aschelminthes are unsegmented animals such as the rotifers, roundworms and hairworms. They all have an alimentary canal with a mouth and anus.

B **Bivalves** are some 8,000 species of flattened MOLLUSCS that consist of 2 rounded, oval or elongated hinged shells in which the soft-bodied animal lives. Large gills filter their food from ocean currents. Many species use their muscular foot to burrow with. Others, such as mussels, attach themselves firmly to rocks with long, sticky threads.

Brittle stars are ECHINODERMS that are star-shaped but have very slender arms radiating from a small central disc-shaped body. They feed on plankton.

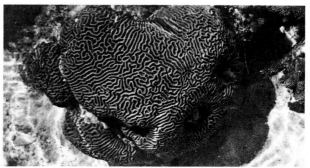

Brain coral

Bristleworms, or ragworms, are marine ANNELIDS that have numerous bristles growing from muscular, fleshy 'paddles' called parapodia.

C **Coelenterates** are the aquatic group that contains the HYDROIDS, JELLYFISH, SEA ANEMONES and CORALS. They are the most primitive of the many-celled animals. Tentacles, placed round the mouth, bear cells that can seize, sting and paralyse prey. There are 2 basic struc-

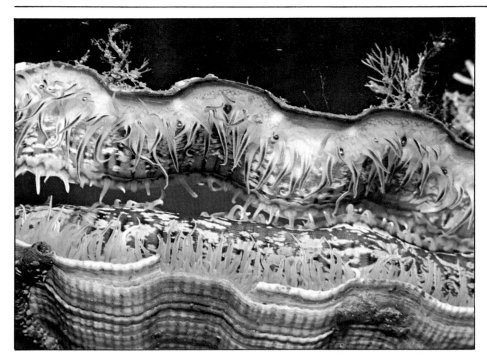

the BIVALVES (two-shelled molluscs) such as scallops and clams. These 'eyes' cannot focus but are very sensitive to light intensity.

Land SNAILS and SLUGS walk on a muscular foot, but the foot has evolved to become flapping wings in the sea butterflies. In some bivalves, such as oysters, the muscular foot is used to drag themselves around but in the scallops, they move by clapping the two halves of the shell together and forcing out the water. As with all bivalves, these molluscs feed on tiny particles suspended in the water or lying on the sea bed. The particles are filtered through the gills of the breathing organs and passed to the mouth.

The echinoderms

The spiny-skinned ECHINODERMS, such as starfish, are often found on the sea shore. These animals have a system of water-filled canals that run through the body. Tiny branches protrude from the skin and are known as tube feet. These are used for movement and respiration. In the starfish, the suction effect of the tube feet is used to open the shells of oysters and mussels. Then the starfish push out their stomachs over the mollusc and secrete digestive stomach juices. The semi-liquid food is sucked up and passed into digestive glands in the arms.

The globular sea urchins are without arms but they also have tube feet in the five rays around

and the lives they lead. Most of the primitive protozoans push themselves along by vibrating long flagella, or smaller hair-like structures called 'cilia'. AMOEBAS push out a portion of their body, towards which the rest of their body flows. They cannot see but are sensitive to light intensity and chemical changes in their surroundings.

The coelenterates, such as the jellyfish and the SEA ANEMONES, have tentacles that are sensitive to touch. The Portuguese man-o'-war, for example, has tentacles and POLYPS which grow to about two metres long. A fish swimming into this dangling mass is almost immediately stunned by the stinging cells and drawn towards the feeding polyps.

The senses of touch and sight are well developed in some species of molluscs. Land snails have two pairs of tentacles on their heads. The smaller, front pair, are thought to be concerned with the sense of smell and the larger pair carry eyes on the tips. Water snails tend to have only one pair of tentacles and the eyes are at the base. The most highly evolved and intelligent molluscs are the SQUIDS, CUTTLEFISH, and OCTOPUSES. The eight-armed octopus has excellent eyesight comparable to that of a vertebrate. Other much simpler eyes (or 'ocelli') are found in

Above: Around the mantle edge of the scallop are located numerous bright blue 'eyes'. Each 'eye' has a lens, cornea and retina but probably cannot see an image. It can detect sudden changes in light intensity.

Right: An octopus swims by rapidly expelling water from the mantle cavity through a funnel.

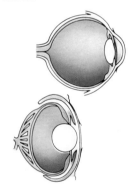

Above: An octopus's eye (bottom) focuses by changing the length between lens and retina, not by changing the lens shape as in a human eye (top).

---

tures: the cylindrical POLYPS, such as anemones, hydroids and coral, attached to rocks, and the free-swimming jellyfish.

Corals, closely related to SEA ANEMONES, are COELENTERATES growing in colonies. Their POLYPS are protected by a hard external skeleton composed of calcium carbonate. Coral reefs are formed by millions of these polyps continuously budding.

Cuttlefish are molluscs with both eyes and an internal shell – the cuttlebone. Like their relatives the SQUIDS and OCTOPUSES, they have an

ink sac near the anus. When alarmed they release the ink to form a dense smokescreen to enable them to escape.

Cuttlefish

E Earthworms range in length from just a few centimetres to over 3 metres. Although they are ANNELIDS, they have just a few bristles and lack parapodia (feet-like projections). They have both male and female sex organs, but certain mechanisms ensure that cross-fertilization takes place.

Echinodermata are marine animals such as the radially symmetrical STARFISH, BRITTLE STARS, sea urchins, SEA CUCUMBERS and feather stars. Most adults can move by means of tube feet, but only

very slowly. The spiny skin gives some protection.

F Flatworms are members of the group PLATYHELMINTHES, the majority of which are parasitic, such as flukes and tapeworms. Flatworms are aquatic animals. Most species are less than 2 cm long. They show remarkable powers of regeneration when body parts are lost.

H Hemichordata are marine animals such as acorn worms. They are the most primitive of the chor-

dates as they possess a notochord (the forerunner to the backbone).

Hydroids are some 2,700 species of COELENTERATES including *Hydra* and *Obelia*. They usually alternate between a POLYP and a free-swimming stage during their life cycle. Many of the marine growths on rock and shells are hydroid colonies.

J Jellyfish, of which there are some 200 species, spend the major part of their lives free-swimming in the oceans of the world. They swim by pulsating their

the body. Their shell is rigid due to skeletal plates but there are holes through which the tube feet pass. The sea urchin is equipped with five strong beak-like teeth worked by a series of muscles. This equipment is used to chew food such as seaweed or dead sea creatures.

The annelids

The ANNELIDS, or ringed worms, live in all types of soil except acidic sandy ones. They are valuable assets to gardeners and farmers because they feed on decaying plant and animal matter, and their tunnelling helps soil cultivation by allowing air and water to reach the soil. Charles DARWIN (*see page 4*) studied earthworms and estimated that every particle of topsoil goes through a worm at least once every few years.

Most bristleworms live in mud or sand burrows on the ocean bed, filtering particles from the ocean currents using their long tentacles. The ragworms are wanderers, using their side feet

known as 'parapodia' to move over the sea bed or swim along. The ragworm has two horny jaws for gripping prey.

The leeches are parasitic annelids. Some feed on small animals while others just suck blood and other juices from larger animals. At one end a leech is usually equipped with a large sucker with which it attaches itself to a host. The mouth opens in the middle of the sucker and some species have teeth.

Corals – the reef builders

The beautiful coral reefs that are found in warm, shallow tropical seas are the results of the work of millions of tiny relatives of the SEA ANEMONES, the corals. Each animal varies from a millimetre to about two centimetres in length and forms a limestone skeleton around its anemone-like body. These limestone skeletons form the basis of the coral reef. The many different coral species form their own shapes such as brain coral.

Below: This mollusc, a bivalve, looks rather immobile when resting on the bottom of the sea bed. However, if a predatory starfish approaches, the scallop takes quick evasive action to avoid being eaten. It moves by snapping its 2 shells together, thus expelling water.

Below: A Portuguese man-o'-war has here trapped and killed a fish in its mass of stinging tentacles.

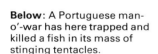

Below: Most spectacular and beautiful of all molluscs, sea slugs breathe through their colourful feathery projections, the cerata, growing from the dorsal surface. The species illustrated is found in the tropical coral reefs off Mozambique.

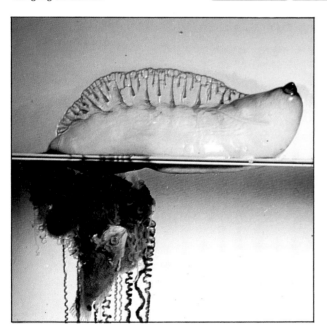

bell and this can range in size from a few millimetres to 2 metres across.

L Leeches are some 300 species of parasitic AN-NELIDS that have few seg-

Pelagia colorata (jellyfish)

ments, no bristles or para-podia and a much reduced body cavity. Most species are aquatic and cling to their plant or animal hosts by suckers which are found at both ends of the body.

M Mollusca are animals with unsegmented bodies, such as SNAILS, BIVALVES, and SQUIDS, and highly developed blood and nervous systems. They do not have a standard shape, the body outline depending on the environment in which the species lives. Most mol-luscs, except the highly-

evolved OCTOPUS group, are slow movers.

O Octopuses are fast moving, highly-evolved

Octopus

molluscs, closely related to squids and cuttlefish. They possess a very efficient nerv-ous system and eyes. The mollusc shell has been lost during the course of evolu-tion. It moves by either pul-ling itself over the rocks by 8 suckered tentacles or arms, or by forcibly expelling water from its funnel.

P Platyhelminthes are the FLATWORMS, flukes and tapeworms, a large group of some 5,500 species.
Polyp is the adult individual of some multicellular organ-

isms such as HYDROIDS. It is either attached by its base to a solid anchorage or forms part of a floating colony.
Porifera are some 5,000 species of SPONGES. They are mainly marine, only a small minority being freshwater-dwelling and they are able to reproduce sexually and asexually.
Protozoa is the large group of over 30,000 species of simple single-celled ani-mals, although this cell is often highly specialized. Most species are micro-scopic such as AMOEBA and the parasite, plasmodium.

Reproduction

In the invertebrate world there are many ways of producing young. An AMOEBA, for example, can reproduce by simply splitting into two, a division termed 'binary fission'. In coelenterates such as *Obelia, Hydra* (HYDROIDS) or SEA ANEMONES, new individuals can be produced by budding. A few cells of the parent separate off and form a bud (a perfect miniature of the adult). This new bud then breaks away to grow into a mature adult animal.

Some lower forms of life reproduce by fragmentation. This occurs where an animal's body breaks up into two or more parts, as in some species of aquatic worm. In part of the jellyfish life cycle, a larva that has been produced sexually by swimming adults is released, settles on a rock, and grows into a sea-anemone-like animal. This grows up in layers, each layer becoming an eight-armed bud, called an

Above: Although snails have both male and female sex organs, they mate and cross fertilize one another. Prior to copulation, a pair of snails indulge in a very peculiar display. When the snails are side on to one another they drive calcareous darts, often called 'love-darts', into the body wall of the other. This somehow stimulates them into mating.

'ephyra', which eventually breaks off and develops into an adult jellyfish.

The majority of higher invertebrates reproduce sexually although there is a great deal of wastage and many die in their early stages. Sexual reproduction takes place in some animals that can also reproduce asexually, for example the *Hydra*. A simple type of sexual reproduction occurs in many single-celled protozoans. Here two individuals fuse side by side and exchange material from their nuclei. However, higher invertebrates usually have male and female sex cells, although one individual may have both. Snails and earthworms, for example, possess both male and female sex cells. The eggs of a pair of mating earthworms, for example, are each fertilized by the other's male cells. Starfish shed their eggs and sperm into the sea-water in huge numbers. Fertilization thus takes place very much by chance.

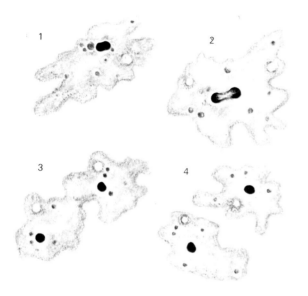

Above: Many lower forms of life can increase their numbers by simply dividing into 2. This process is called asexual reproduction. The nucleus in *Amoeba* divides (1) with the daughter nucleii moving apart (2) as the rest of the cell separates and forms 2 daughter cells (3). The young adult protozoans (4) can repeat the process in 3-4 days.

Right: New young are produced in *Hydra* (seen here) and sea anemones by budding. A bulge grows out from the parent and develops into a tiny replica of the adult. It then breaks off.

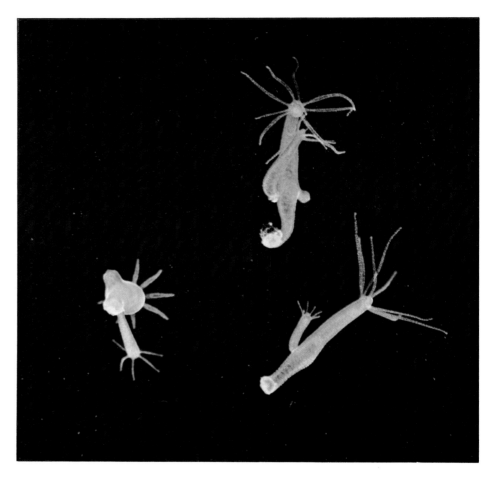

R **Ribbonworms** are some 750 species of the phylum Nemertina, related to the FLATWORM. Some are more than 20 metres long. Most species live in shallow waters.

Roundworms are some 10,000 species of worms with long bodies, pointed both ends and covered by a thick horny cuticle. They may be parasitic or free living, ranging in length from microscopic to over 1 metre.

S **Sea anemones** are related to the CORALS and do not have a free-swimming stage in their life cycle as do the JELLYFISH. They are amongst the most familiar of the COELENTERATES because they are easily seen in tidal rock pools.

Sea cucumbers are some 900 species of elongated, armless ECHINODERMS. Tube feet near the mouth are adapted as tentacles and trap small animals.

Snails and slugs are gastropods, part of the MOLLUSCA phylum. Snails have a single shell whereas slugs are shell-less. A typical snail shell is a conical shape with spiral markings.

Sponges of the group PORIFERA are classified into 3 types. The calcareous sponges have a support of chalk (calcium carbonate). They are either straight or with 3 or 4 branches. Glass sponges have their skeleton composed of hard bits of silica, each having 6 branches. These often fuse to give a lattice structure. The horny sponges have a jelly-like substance between the cells and a skeleton which is a combination of silica and a horny substance called spongin.

Squids are close relatives of CUTTLEFISH and OCTOPUSES. Like the former, they have 8 short and 2 long arms. The long arms wrap up prey after the powerful beak-like jaws have severed the nerve cord

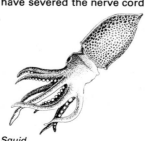

Squid

of a fish. They are able to change their colour and patterning to match surroundings. The giant squid of the North Atlantic is said to reach 20 metres in length.

Starfish and BRITTLE STARS are some 4,500 species of ECHINODERMS. Most have 5 arms radiating from the central disc, though some have as many as 40. They have remarkable powers of regeneration. Provided 20% of the central disc is attached to an arm, an entirely new adult starfish will slowly grow. Most species are carnivorous.

Ecology is the study of animals and plants in their natural habitats. By studying an animal as part of an animal and plant community, we can better understand the inter-relationships which hold together the delicate balance of nature.

Adaptation to Environment

Deserts: Fennec fox

Temperate forest: Red deer

Tropical forest: Toucan

Mountains: Mountain goat

Polar regions: Seal

Right: The major biomes of the Earth are climatic zones each supporting its own specially adapted plants and animals.

Ecologists – scientists that study plants and animals in their natural habitat – have found that each region has its own kinds of plants and animals adapted to its particular climate and surroundings. These animal and plant communities are called "biomes" and the map shows the world's major biomes. Ecologists do not always agree on the exact number of biomes found in the world, some identifying a far greater number than shown here. However, it is generally accepted there are nine major land biomes: the polar regions, the TUNDRA, the CONIFEROUS FORESTS, the DECIDUOUS FORESTS, the TROPICAL FORESTS, TEMPERATE GRASSLANDS, TROPICAL GRASSLANDS, the deserts and the dry scrublands. The waters of the world are usually divided into the fresh waters, the seashores and the oceans.

The succession of vegetation from the equator to the two poles follows a certain pattern. At the equator we find tropical forest, next deciduous forest, then coniferous forest. Then we find grasslands and these areas merge polewards into the area of mosses and lichens, often called the tundra. The same pattern of succession in vegetation is found from the base of a tropical mountain to its peak as the climate changes with the increase in altitude.

Animal adaptation to its environment

The shape and size of an animal's body is usually adapted to its natural surroundings. A warm-blooded bird or mammal has to ensure that its body temperature remains constant. Adequate insulation in the form of fur, feathers or fat is therefore essential. The larger the animal, the more slowly it loses heat. So the Arctic POLAR BEAR is twice the weight and size (2.5 metres in length and half a tonne in weight) of the sun bear that lives in the tropical forests of south-east Asia. Also the polar animals have much thicker fur than their tropical relatives in order to

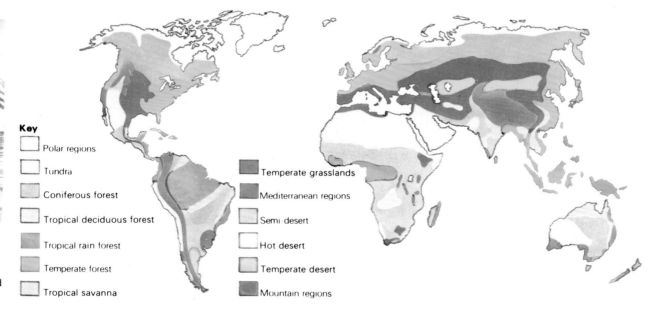

Key

- Polar regions
- Tundra
- Coniferous forest
- Tropical deciduous forest
- Tropical rain forest
- Temperate forest
- Tropical savanna
- Temperate grasslands
- Mediterranean regions
- Semi-desert
- Hot desert
- Temperate desert
- Mountain regions

Reference

A **Adaptation** is a characteristic that aids the ability of a plant or animal to cope with its environment. **Adaptive radiation** is the evolutionary diversification into a variety of ecological roles of species which all have the same common ancestor. For example, the 12 species of finch on the Galapagos Islands evolved from a seed-eating ground finch from South America. Each species is adapted to a particular niche. One finch feeds on cactus, another on seeds, and one species uses a twig to poke out grubs. **Antarctica** is the south polar region, the large frozen continent which sprawls over the 'bottom' of our planet. Apart from a few invertebrates, the only animals that have adapted to the Antarctic conditions are seals, penguins and some sea birds. These are all carnivores, feeding on the rich plankton and fish stocks of the Antarctic seas. **Arctic** is the north polar region, the Arctic Ocean and its surrounding ring of land. The only non-migratory mammals to be found are polar bears, Arctic foxes and members of the seal family.

B **Blind cave fishes** have lost their power of full sight due to becoming adapted for a life in the total darkness of cave pools. They

School of dolphins

find their way about without bumping into anything by using their sensitive lateral line system (*see page 38*).

C **Camels** are adapted for DESERT life with broad, heavy feet that do not sink into sand, nostrils that can be closed to keep out flying sand, and interlocking eyelashes that protect the eyes against both Sun and sand. They can go for considerable distances without water, drawing on the fat reserve in their humps for an energy source.

Chromatophores are

Right: Plants and trees in a jungle are stratified, or layered, and different kinds of animals live in more or less restricted vertical ranges. Each layer has its own characteristics. Animals that live in south-east Asian forests are shown here.

insulate the body, keeping heat in and cold out.

Cold-blooded animals – the invertebrates, fishes, amphibians and reptiles – work more efficiently when warm, but must gain heat from their environment as they are unable to produce heat themselves. Few of these species are found in cold climates, the majority being adapted for warm and tropical biomes.

Tropical forests

Animals living in tropical rain forests, often called jungles, have many problems. Quick movement is rather difficult where these tall trees grow quite close together and are laden with hanging vines and lianas. As a result the animals living there, either blend in with their surroundings as they hide from enemies, or they are well adapted for moving through the branches of the trees. Monkeys, with their gripping hands and balancing or 'prehensile' tails, are very agile among the middle layers of the forest canopy. Certain apes, such as the small gibbons, chimpanzees and orang-utans, all have long arms for swinging from branch to branch. Sloths have adapted to hang upside down and in their South American home stay on one tree as long as it has enough foliage for them to eat. Their fur is encrusted with a green algae which acts as a camouflage among the leaves. Reptiles and amphibians abound, together with tree snakes, tree frogs and numerous lizards. This biome is a haven for birds, which tend to have vivid colours. At a distance, a brightly-coloured bird on a branch could be taken for a tropical flower and it is not until it flies that its identity is revealed. Insects are numerous in the jungle, each one adapted to a particular niche.

Temperate forests

Evergreen forests are made up of conifers such as spruces, firs, pines and hemlock. They are found to the south of the moss and lichen tundra across Eurasia, Alaska and Canada. Many animals of these regions are adapted to eating parts of the conifers. Birds such as redpolls and crossbills feed on the seeds of the cones. Woodpeckers

Labels in illustration: Slow loris, Flying squirrel, Giant hornbill, Flying lizard, Pit viper, Tree frog, Gibbon, Swallowtail butterfly, Leopard, Tree shrew, Malayan moon rat, Gaur, Tiger, Pangolin

---

specialized cells in the skins of certain animals, such as flat fishes, reptiles, amphibians and octopuses, that contain various pigments. Each chromatophore is bounded by an elastic membrane and its size and shape

Camels in the Libyan Desert

is controlled by special muscles. When light intensity in the surroundings alters, a message is passed through the eye, via the nervous system to the chromatophores which expand (spreading more pigment) or contract (spreading less pigment) to blend the animal's skin to the new environment.

Climbing perch were so named because when the first ones were found in trees it was thought they climbed there unaided. Later it was realized that they had been dropped there by crows or kites. However, they do leave the water to move from pond to pond, using their strong gill covers like arms and pushing with their pectoral fins and tail. They have air-breathing chambers connected to their normal gills and must have atmospheric air to survive.

Colour change in animals is usually brought about by CHROMATOPHORES. Other changes occur as part of the process of becoming an adult. For example, a gull loses its brown juvenile plumage by moulting and growing new white feathers.

Other birds and mammals, such as PTARMIGANS and Arctic foxes, change their coats gradually to match the different seasons. These animals change from brown to white in autumn, thus making them well camouflaged for Arctic winters. Many birds grow bright coloured feathers for the courtship season, the more vivid decorations being usually worn by the males.

Coniferous forests are characterized by long, cold winters and short warm summers. The animals of this type of biome are often

Left: The great diving beetle is adapted for an aquatic life, although it still breathes air. It pushes its abdominal tip just above the surface, raises its wing covers and draws air into a pair of breathing pores (spiracles). It also traps air under its wing covers as a reserve.

extract grubs and insects from the bark. Mammals such as squirrels also take the seeds as part of their diet. Insects abound in coniferous forests; often one species has adapted to feed on a particular species of conifer.

In deciduous forests there is a mixture of trees such as oak, birch, beech and ash. These trees shed their leaves in the autumn, and so there is more light available for animals living on the ground. More plants are able to grow on these forest floors than in the jungle and coniferous forests, so that a greater number of animals inhabit this area. These include well-known animals such as foxes, badgers, squirrels, deer, and ground-nesting birds such as pheasant and woodcock. The canopy layer is home for many birds such as woodpeckers, warblers, tits, owls and buzzards. Each bird species is adapted to its own type of feeding, so that there is little competition.

The polar biome
There is little vegetation within the polar regions. Without brief periods of warmth, only the simplest plants can grow. Most animals are therefore flesh-eaters, such as the polar bears and Arctic foxes of the northern lands, and the fish-eating penguins of the Antarctic. In the northern polar region most of the animals feed on the tundra during the short summer and then move south when the snows come again. The musk-oxen, snowy owls, ptarmigan and snow buntings are some of the few animals that can remain on the tundra all the year round.

The grassland biomes
In the areas where there are very few trees and stretches of open grassland, the animals must be able to move quickly as there is so little cover. There are large areas of grassland on all the continents. The tropical grasslands are called savannas, while in the cooler parts of the world they are termed prairies, plains or steppes.

The grasslands of temperate regions were formerly very large, but most areas in the Northern Hemisphere have been ploughed up for

Right: The woodpecker is so named because it chips holes in trees to prey on insects or to make a nesting hole. When a hole is drilled, the long flexible tongue then darts forwards and its tiny barbs and sticky saliva at the tip catch the prey. The tongue can extend 4 times the length of the bird's upper beak and runs back inside its head.

---

climbers or adapted for a life on the forest floor. Common climbers include tree squirrels. On the forest floor roam

Red squirrel

wolves, bobcats, and black and brown bears. Insects are abundant in summer and provide a food supply for the many bird species.
Convergent evolution occurs when similar animal characteristics are developed over long periods of time in species that live in the same kind of environment but in different areas of the world. For example, the toucans of the South American tropical forest look very similar to the hornbills of African and Asian tropical forests, although they are not closely related.

Termite mounds

D **Deciduous forests** are made up of a mixed variety of trees that need warm summers, mild winters, and a moderate rainfall well spread out over the year. Insects abound in summer and include butterflies and moths and their caterpillars, bugs and

beetles.
Deserts are very dry areas where days are very hot and nights are quite cold. As the Sun sets, the temperatures can fall from a sizzling 56°C to 0°C. Deserts cover almost 20% of the Earth's surface and are not as lifeless as many people suppose. The adaptation of animals in this biome is concerned mainly with getting enough water and to avoid burning up in the middle of the day. Special adaptations include longer legs to keep away from the burning sands or the ability to fly onto plants.

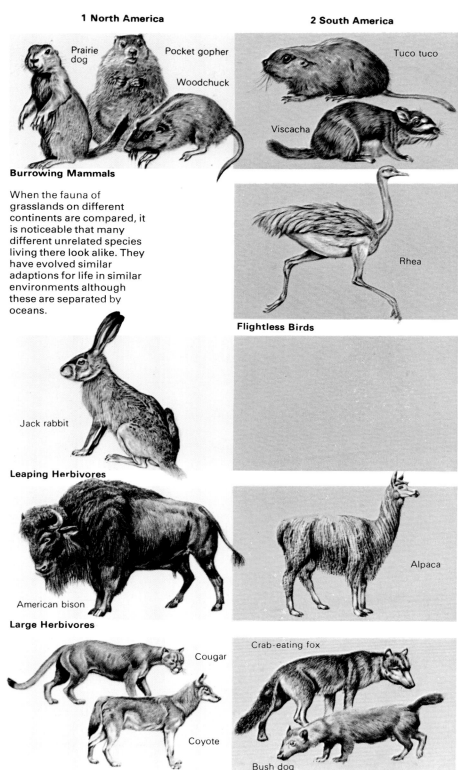

1 North America

Prairie dog

Pocket gopher

Woodchuck

2 South America

Tuco tuco

Viscacha

Burrowing Mammals

When the fauna of grasslands on different continents are compared, it is noticeable that many different unrelated species living there look alike. They have evolved similar adaptions for life in similar environments although these are separated by oceans.

Rhea

Flightless Birds

Jack rabbit

Leaping Herbivores

American bison

Large Herbivores

Alpaca

Cougar

Coyote

Crab-eating fox

Bush dog

Large Carnivores

agriculture, grazed by cattle, or built on. In the Southern Hemisphere, domestic herds have become dominant over the natural animal life. It is very interesting that the same environment around the world has produced similar looking animals, although they are not closely related. This is because as the various animal species evolved in the grassland environment of the different continents, and as the conditions were predominantly the same in each, they ended up with the same adaptations and thus looked very similar. This is known as 'CONVERGENT EVOLU-TION' and the chart on the right shows many examples.

Dry lands and deserts

Where the rainfall is too low to support the tropical grasslands, there is a gradual change through temperate grasslands to scrub, to semi-desert and then to dry arid · deserts. The scrublands have their own unique animals. In Australia there are some curious marsupials (pouched mammals). The koala lives on a diet of eucalyptus leaves alone. The FENNEC FOX is adapted to living in the desert. Its huge ears help to keep it cool. Most desert animals have to avoid the heat by burrowing or hiding during the day. Many obtain moisture from desert plants, as water is very scarce. Thus, the majority of these animals are nocturnal, moving and feeding during the cooler night. Many cold-blooded reptiles bask in the Sun's heat in these areas in order to raise their body temperature to a level suitable for activity.

Strange relationships

Within each biome there are various small-scale habitats, such as the life in a tree, a fallen log or a rock pool. Within each habitat, certain animal species are adapted to a certain way of life. The place where an animal lives and its behaviour

---

Predators drink blood from their prey, while herbivores 'drink' water from desert plants such as cacti.

F **Fennec foxes** are adapted for DESERT life. Their pale short coats and huge ears pick up the slightest sound in the hours of darkness when they hunt for food. The ears also help to cool them when their bodies get overheated.
Fresh water biome in- cludes the temporary rain- pools of desert regions, ponds and streams to an enormous inland mass of

water, such as the Caspian Sea. The oxygen content varies from one part of a lake to another. This can be caused by the type of plant life, currents or lake-bottom material. In a rapid stream there is lots of oxygen in the bubbling, rushing waters but the animal must be able to cope with the strong cur- rents.

G **Grasslands** cover near- ly 50% of the land area in the Southern Hemisphere. There are also great areas of flat, open country in the Northern Hemisphere.

Jackal

Speed is the first important adaptation of large mam- mals such as antelopes, zebra and deer. This enables them to escape from enemies such as cheetahs, lions and packs of jackals, all good running carnivores. Many small mammals have adapted to jumping along instead of running, using long strong back legs and balancing tails. Even some birds have evolved into run- ners, from the flightless os- triches and emus to the long-tailed roadrunners of the south-west United States.

H **Hermit crabs** are more closely related to lobs- ters and shrimps than true crabs. The long tail has become adapted to hold the hermit to the inside of an empty shell. When dis- turbed, the hermit draws back into its shelter leaving only its pincer claws show- ing. They change shells at frequent intervals to keep up with their growth. They often provide a moveable home to certain species of sea anemone. This is not a chance association. The anemone gives the crab camouflage while the

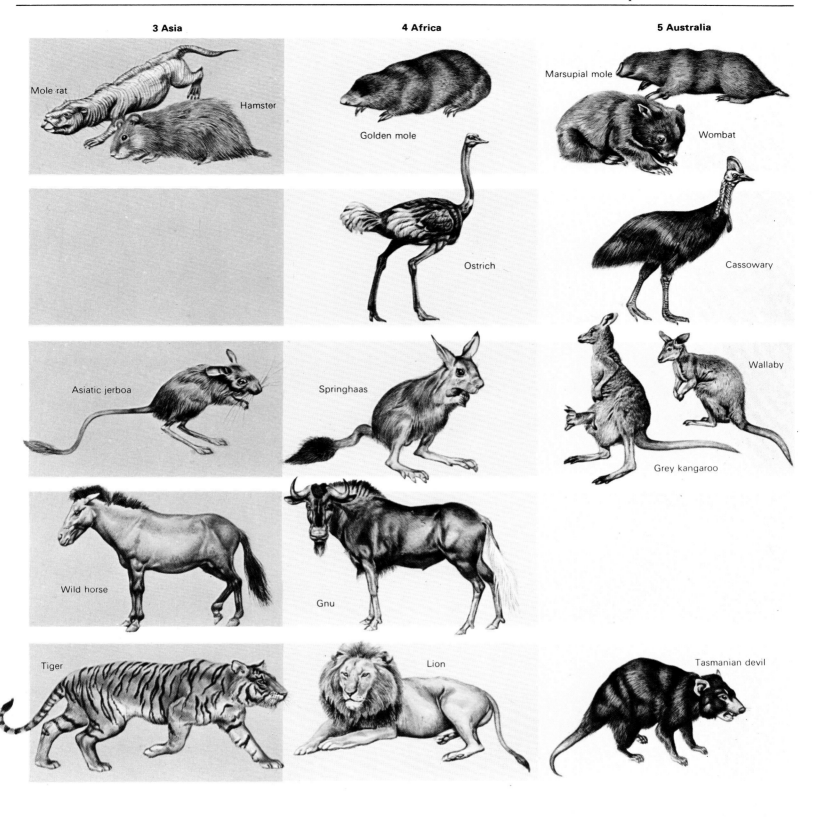

3 Asia

Mole rat

Hamster

Asiatic jerboa

Wild horse

Tiger

4 Africa

Golden mole

Ostrich

Springhaas

Gnu

Lion

5 Australia

Marsupial mole

Wombat

Cassowary

Wallaby

Grey kangaroo

Tasmanian devil

anemone is transported to new feeding grounds.

J **Jacanas,** or lily trotters, are tropical birds with

Hermit crab

very long toes and straight claws. These enable them to walk easily over floating vegetation without sinking.

M **Mimicry** is where an animal resembles some other animal, plant or even object in its environment. It is used to deceive predators or prey and can confuse the creature with the object that it mimics. Stick insects, for example, look like the twigs on which they live. False coral snakes are not poisonous but their red, black and white bands down their bodies mimic the

venomous coral snakes.

N **Nocturnal creatures** are those animals that are active during the hours of darkness (from dusk to dawn). Nocturnal primates, such as the tarsier and loris, have huge eyes (necessary to achieve a high degree of stereoscopic vision) and excellent hearing so that they can detect both food and enemies, and find their way about the branches of their tree home.

O **Ocean biome** is the term given to the huge

marine zone. The animals and plants usually live and move in one of 4 different ways. Some float at the surface, travelling from place to place using the currents and surface winds. Others are adapted to swimming and include most fishes, seals, whales and penguins. Others crawl on the ocean floor, for example worms, starfish and molluscs. There are also the deep-sea dwellers which have adapted in bizarre ways to ensure survival. The female angler fish ensures that a mate is available for

reproducing by having a dwarf parasite male angler fish attached to her body.

Okapis of the Congo TROPICAL FORESTS exhibit disruptive coloration. This breaks up the animal's body outline and makes this shy animal more difficult to see by a predator.

Oxpeckers are African starlings that feed exclusively on ticks that live on mammals such as antelopes, rhinos and hippos.

P **Pilot fishes** are so named because of their peculiar habit of appearing

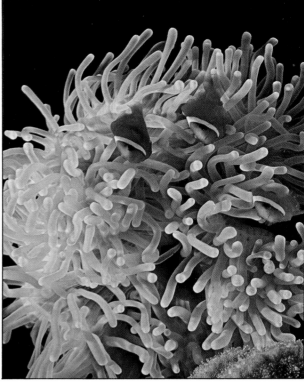

there is called its niche. For example, the niche of an adult barnacle is being attached to a rock, feeding from the water current when the tide is in, and closing its shells when exposed to the air or to the flow of the outgoing tide.

In certain habitats animals have entered into some strange relationships. One well-known partnership is that of the colourful tropical anemone fishes and certain tropical sea anemones. These tiny fishes (some called clown fishes are illustrated *above*) live in and around the anemone's stinging tentacles, being curiously immune to its lethal barbs. The body of these fishes is believed to be covered with a mucus or slime secreted by certain glands of the fish, which inhibits the action of the stinging cells. In return for this protection, it tempts other fish into the anemone's tentacles.

Anyone who has been fortunate enough to visit an African game reserve will probably have seen egrets travelling upon the backs of cattle, elephants or antelopes. These birds catch the insects that are disturbed by these large herbivorous mammals. Although Africa is the original home of these cattle egrets, they have

Above: The remora is perfectly adapted for hitching rides on fishes. Its dorsal fin acts as a sucker and attaches the remora to its host's body.

Below: Red-billed oxpeckers feed on ticks off their Brahman cattle host and in return give warning calls when any danger threatens.

Above: Red clown fish swim fearlessly among the sea anemone's stinging tentacles. The fish are protected by a mucous covering which inhibits the action of the stinging cells.

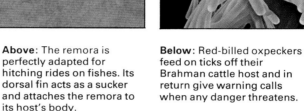

to direct the course of sharks and other large fishes. It is unlikely that they really do this, but they do benefit from the food scraps of the shark's leftovers and to some extent are protected by them.

Polar bears are adapted for the extremes of Arctic life with thick white coats, hairy soles on their feet and small ears that do not lose too much heat to the outside.

Ptarmigan are small grouse of the Arctic tundra and brushlands. They are unique in having their toes and legs completely feathered, sup-

posedly an adaptation for walking in soft snow. They live on berries, buds, seeds and lichens. They are typically grey-brown in summer and white in winter.

Red grouse and young

S Scrubland is a fringe biome between hot deserts and tropical grasslands, sometimes referred to as semi-desert. The rainfall is very irregular and when it does come, plant life bursts into bloom and lower animals such as insects reproduce rapidly under the improved conditions. There are large stretches of scrub in Australia, this mallee scrub providing a habitat for many birds such as bell birds and mallee fowls. These fowl incubate their eggs in a 'compost' of rotting vegetation.

Seashore is the biome where land meets the ocean and is the area between the high and low tide marks. The rich variety of animal life living between the tides is exposed to sharper contrasts of environment than any other living creatures. They must withstand wave action when the tide is in, and drying out by the wind and Sun when the tide is out.

Sloths are members of the Edentata order that are renowned for their slow movements. A sloth clearly in a hurry was timed at 4 metres a minute. The sloth is

been introduced by man into other countries of the world, such as America and Australia, where they have been seen following tractors to catch the disturbed insects.

Some of the most curiously shaped insects are the 1,800 species of praying mantis. The various species are perfectly coloured and often shaped to match their background. They can sit motionless, totally camouflaged, waiting for an innocent fly or other insect to come within reach. Then the folded front limbs shoot out to grasp the victim.

As in the mantis, it is often the way an animal feeds that has produced some strange adaptation. Chameleons are slow-movers but perfectly camouflaged in their African and Madagascan tree homes. Their limbs do not shoot out but the muscular tongue is the organ which shoots out and catches prey. Vampire bats have evolved to feed only on blood. They silently alight at night near a sleeping victim, such as a domestic cow or giant ant eater, and climb up gently to bite a piece of skin from a relatively hairless area such as the neck region. The victim's blood does not clot as the bat feeds due to an anti-coagulant in the bat's saliva.

Above: A camouflaged female praying mantis waits motionless pretending to be a twig, until an insect passes by. It then suddenly jack-knifes its front legs out to grasp the prey. Mating is a hazardous business for the male mantis as he is liable to end up being eaten as well.

Left: A chameleon changes its colour in response to light intensity and this often results in a better blend with its background.

adapted to a life of hanging upside down in trees. Its hair grows backwards from the belly towards the back to prevent the fur becoming waterlogged. Its coat has a greenish tinge from the algae that live in the fur and this aids camouflage.

Taiga is the term given to the vast stretches of evergreen forests that grow right across northern Europe, Asia and northern America, just south of the tundra.
Temperate grasslands was the term formerly used

Swallow

for the Great Plains or prairies of North America and the vast steppes of Eurasia. Today, most of these areas are farmland,

although the untamed and untouched regions still support a unique range of wildlife. Before the settlers came, huge herds of bison

Red fox

Blend and bluff camouflage

We have seen numerous examples of animals that are camouflaged to blend in with their surroundings and avoid predators. If they keep still, an observer finds it very difficult to detect them. There are thousands of insects looking like leaves, green grass, flowers or even sticks, these various disguises helping them to survive longer and reach their goal of producing young and ensuring the survival of their kind. Many ground-nesting birds are camouflaged so that they are hidden when on their nest and incubating the young.

Few mammals draw attention to themselves by being brightly coloured, and most of the young are very well camouflaged indeed. This is because most mammalian young are unable to run and feed when born, and must remain hidden until strong enough to accompany their parents. Young fawns are usually spotted when they are born in order to blend in with their patchy background. At this stage they give off no scent either, so that a hunting fox or bear will not be able to find them.

An animal's camouflage pattern is not always fixed. Some animals can change colour and blend with their background. This is mainly true of fishes and includes plaice and halibut and other flat fishes, as well as octopuses and some other molluscs. On land lizards, such as chameleons and anole lizards can change colour. The reason for this occurrence is due to changes in light intensity in the surroundings.

Warning colours

Some insects, snakes, fishes and frogs, and one or two lizards, are poisonous or distasteful to eat, and these are usually brightly coloured. This warns other animals that they are to be left alone. However, the predators have usually had to learn this fact by trial and error. For example, wasps, cinnabars and their caterpillars are bold yellow or red and black. It has been shown that predators quickly associate the bright colours with a bad taste and leave them alone. Another strange adaptation is where harmless insects and snakes have evolved similar colours to the poisonous ones. They gain protection from mimicking the poisonous species, but the extent of this protection is open to a good deal of argument.

Above: The bright colours of an arrowpoison frog from the jungles of South America warn predators of its deadly venom. The poison is used on Indian arrows and can paralyze animals as large as monkeys.

Below: The eyespots on a peacock butterfly provide it with effective protection. They may scare away enemies, such as small birds.

---

and pronghorn antelope roamed the prairies. A large rodent population is adapted to prairie and steppe life and includes prairie dogs, ground squirrels and gophers.

Tropical forests are the zones we often call jungles. The constantly warm temperatures and frequent rainfall are very important to the diversity of animal life. The giant trees tower over 60 metres high and animals are adapted to life in the various layers from the tree tops to the jungle floor. Rivers are plentiful in this environment

Indian elephant

and many animals, for example tapirs, peccaries and capybaras (both South American rodents) will take to water to avoid being caught by predators such as jaguars.

Tropical grasslands are often called savannas and large areas are found in Africa, as well as South America and Australia. These grasslands have dry seasons and the animals living here must be able to survive periods of drought. The African savanna is extremely rich in animal life, many of the mammalian species being quite large. These include elephants, zebra, gnu and buffalo as well as many kinds of antelopes and gazelles. There are also many predators

such as lions, cheetahs and hyenas and jackals.

Tundra is the inhospitable biome that runs around the edges of the ice-covered Arctic. The lower layers of the tundra earth are permanently frozen and are called permafrost. There is a 4 month summer during which time plant life flourishes, insects abound and these support a large influx of birds such as waterfowl, finches, buntings, larks and warblers.

Throughout history, many species have become extinct as better adapted species replaced them. But man has created a crisis in the natural world. Today, we are just beginning to understand the dangers of upsetting the balance of nature.

Man and Animals

Man, *Homo sapiens,* has existed on the Earth for only about 300,000 years. During this time he has used and been useful to many other animals. Certain parasites such as fleas and bugs may live on his body; others such as tapeworms and flukes can live inside his body. Some animals such as crocodiles, tigers and lions may kill him, while man has hunted birds, mammals and fish for thousands of years for food and clothing.

An interesting aspect of man's involvement with animals, is the fact that he has been able to tame several species to work for him. The domestication of animals has been going on for over 10,000 years and probably wolves and jackals were the first animals to be domesticated. In the Stone Age, man was still a nomadic hunter and the wolf existed over a wide area throughout Europe and Asia, down to peninsular India, and in North America. Man and the wolf, therefore, often came into contact. The earliest domestica-

Above: Dogs are popularly called man's best friend.

tion of dogs probably took place in south-west Asia and when man reached Australasia in the Middle Stone Age he took some with him. Some escaped and returned to a wild state and these are the DINGOS we know today. Wolves and dogs probably first associated with man to feed off the remains from the kills of early tribesmen. Possibly men reared puppies they found, at first keeping them as sources of food when times were hard. No doubt men realized their worth as watch dogs and later used them to seek out prey. They were also trained to herd and protect other domestic animals such as goats and sheep. By the time the dog reached Europe, about 6000 BC, it was probably already performing the duties of a sheepdog.

Today there are over 165 BREEDS of dog which have been bred for racing, coursing, carrying, guiding, herding, retrieving, guarding as well as just for being a faithful friend for man. There is

Above: Horse-riding is a popular sport throughout the world.

Above: South Americans use llamas on rocky terrain.

Above: Indian elephants are trained to work in forests.

Above: Camels are widely used in dry climates.

Above: The Lapps depend on reindeer for their livelihood.

Reference

A **Aye ayes** are cat-sized and do not look like primates at all, although their nearest relatives are the lemur-like indris. They are nocturnal creatures, living in the tropical forests of Madagascar. They have claw-like nails on the 5 digits of each foot. They have rodents' incisors to bite holes into the bark and the extra-long middle digit of the hands extracts any grub. Probably less than 50 exist.

B **Blue whales** are the largest of all mammals but despite being totally protected probably less than 2,000 survive. They were the

Java sparrow

most avidly hunted whales between 1920-40 as they were the biggest and gave the greatest quantity of oil. In 1930 it was estimated between 30-40,000 existed.
Breed is a race or strain of a domesticated animal which continues to breed true only if crossed with breeds having the same hereditary qualities.

D **Dingos** are the wild dogs of Australasia, about the size of a collie, standing about 50 cm at the shoulder. They vary in colour from light red to

brown and yellow-brown. Although rewards were given for killing them as they killed sheep and cattle, they are still common.
Dodos were flightless

Dingo

turkey-sized birds that lived on the island of Mauritius. Sailors hunted the dodo for food, and pigs destroyed its eggs. It had become extinct by 1681.

E **European bison,** also known as the wisent, were almost exterminated in the truly wild state. They once ranged over western and southern Europe, east to the Caucasus and north to Siberia. They were saved just in time by being protected in the 1930s. Today, numbers are healthy, the majority of them living in the

now a breed available to suit almost anyone, wherever he lives and whatever he does.

In domesticating wild animals, man has always selected certain characteristics that he would like to be reduced or accentuated in future stock. For example, from the point of view of domestication, one problem is the fierceness and aggressive nature of most wild animals. This has been reduced to a minimum by selective breeding. The inborn traits of certain domesticated animals such as the CAMELS, reindeer and yaks have not been greatly altered because their value to man is that they are particularly well adapted to their environments. The reindeer has not had its instinct to migrate bred out of its system. Instead, the Laplanders have adapted to the reindeer's life and lead a nomadic life following the semi-domesticated reindeer herds on their journeys. The working elephant of the Asian timber forests has also not been changed, but each new offspring is taught the logging trade during its first 15 years of life, its adolescent period of growth.

Most domesticated animals, however, have been bred to provide meat, milk or skins, or to be draught animals or beasts of burden. The many breeds of pig are all descended from the fierce forest-dwelling wild boar. Today they are strikingly different from their wild ancestors. They are less hairy, shorter-legged, fatter and have quite differently shaped heads. They fatten more easily and reproduce at a greater rate than their wild cousins.

All domestic horse breeds probably originate from the species *Equus caballus*. PRZEWALSKI'S HORSE is the only surviving representative in the wild of the original species. It exists in small numbers on the Mongolian steppes, as well as in captivity. Donkeys, mules and asses have been man's beasts of burden for centuries. They are still used to carry people and goods in poorer countries, but in most developed countries, donkeys are sometimes kept as pets for children.

Sheep, goats and cattle belong to the same family as buffaloes and antelopes. It is not known which wild sheep is the ancestor of modern breeds but the mouflon probably provided the main breeding stock centuries ago. Sheep are valued commercially for their mutton and wool, and in some countries for their milk. Goats are valued either for their fleece or their milk, the

cashmere and the angora breeds being the most important goats, reared for their soft, fine wool.

Domestic cattle are known to have existed in Babylon as early as 5000 BC. In continental Europe, domestic cattle were bred from the wild aurochs, enormous long-horned beasts that stood 1.8 metres at the shoulder. The last known wild specimen was killed in Poland in 1628. A very ancient breed which survives in a few places is the 'wild' white cattle which was probably brought into Europe by the Romans, although some zoologists think it is a direct descendant of the wild auroch. However, today the numerous breeds of domesticated cattle are kept for their milk, meat and hides, and also as beasts of burden in poorer countries.

Some animals are very difficult to breed in captivity and are usually caught when young and then tamed. These include those animals we usually refer to as exotic pets. The ancient Egyptians, for example, kept falcons, cheetahs and mongooses. FALCONRY is a popular sport in many countries today, but has led to the poaching of young protected birds-of-prey from their nests in many countries, so seriously endangering various species such as the peregrine.

Endangered species

Over the last few hundred years many hundreds of species of plants and animals have disappeared from the Earth. Some have become EXTINCT naturally during the course of evolution. However, man has been responsible for much of

Left: The sport of taking a quarry using a bird-of-prey was practised in the East as long ago as 1200 BC. It became the sport of kings and aristocracy in the West during the Middle Ages. A peregrine falcon is seen here ready to take off.

Right: All dogs belong to the same species and wild wolves were probably their original ancestor. Working dogs such as those illustrated are most important to man, but many 'toy' dogs have been bred purely as pets. The earliest domesticated dogs probably resembled the Australian dingo. These are descended from the dogs that Stone Age man took to that continent from Asia about 8,000 years ago.

Right: Domestic pigs originate from the wild boar of Eurasia and south-east Asia. The earliest known domesticated pigs existed in Neolithic times. Selective breeding brought about a less hairy skin and the loss of tusks. Pigs are valued almost solely as meat producers, but the meat is used in different ways. Bacon pigs are longer bodied, rather higher on the legs and a lighter colour than a pork animal of similar age.

Right: The origins of sheep are not known for certain. The view most widely held is that the Asiatic mouflon provided the foundation stock, but the Urial and Argali are sometimes regarded as its ancestors. Domestic sheep are valued for their mutton and wool, and occasionally also for their milk. With some breeds, such as the Merino, wool is the primary product, while other such as the Southdown are kept mainly for their mutton, and wool is a secondary consideration.

Bialowieza Forest in Poland and various zoos. A woodland animal, it browses on ferns, leaves and bark.
Extinct animals are those that have died out completely. This can happen naturally due to changes in the environment to which an animal cannot adapt. Man's pressure on wildlife has altered the rate of extinction so that now many more species are endangered. Animals are hunted for food, sport, pharmaceutical use, or because they are considered vermin or pests. Others have been killed by

Black rhinoceros

various types of pollution and the destruction of their habitat.

F **Falconry** is the hunting sport that uses falcons, such as the peregrine, in open country. In former times it was a sport of gentlemen and kings. It still is widespread in the Middle East. Today poachers take fledglings to train as the bird is very difficult to breed in captivity. Pesticides and other toxic chemicals have also taken their toll on the population. Although protected and no longer in danger of extinction, the peregrine's chances of increasing in number are still uncertain.

G **Game reserves** or game parks have been established in most countries of the world to protect

Leopard

the wildlife living in that region.
Giant pandas were first introduced to the western

Right: The cattalo, half cow and half bison, was bred in the USA in the late 1880s. It did not prove to have the beefier qualities of its longhorn mother and the sturdy and hardy characters of a bison. Sadly, the offspring were stillborn or sterile and those that survived were very bad tempered. The idea soon floundered.

Bison

Longhorn

Cattalo

The **wolf** is probably the ancestor to early breeds such as terriers.

The **husky** was originally bred in Greenland for sledge pulling.

The **alsatian** is widely used as a guard dog and for police work.

The **rough collie** is used for sheep work.

The **red setter** is used as a gun dog for retrieving.

The **wild boar** was the fierce forest ancestor to modern pigs.

The **Berkshire** breed originated in the Thames Valley, England, for meat.

The **Landrace** was selectively bred for bacon by the Danes.

The **middle white sow** was bred from Yorkshire breeds.

The **Tamworth** is bred in the Midlands for high quality bacon.

The **Asiatic mouflon** was probably the original ancestor of the modern-day sheep.

Jacob's sheep was an early breed that is quite rare today.

The **Southdown** is a short-wool breed with excellent meat.

The **Leicester** is the oldest long-wool breed, dating back to the 1700s.

The **Merino ram** is a long-wool breed that is very popular in Australia.

---

world by Père David, the French missionary, in 1869 when he returned from China. These pandas look like bears but they are actually related to the racoons. Their range is restricted to the bamboo forests in an isolated mountainous region of western Szechwan in China. Little is known of their breeding habits in the wild but the Chinese have bred several in captivity and probably the future of this attractive mammal is ensured. Over 20 specimens are in captivity, the majority being in China.

Golden lion marmosets are the most brightly coloured of all living mammals – an intense, shimmering golden yellow. They are found in the coastal forests of south-eastern Brazil where numbers are below 400. Their decline is due to deforestation and the capture of live specimens for the animal trade. Although protected in Brazil it may be too late to save this primate.
Green turtles are marine reptiles of warm seas that come ashore to tropical sandy beaches only to lay their large clutches of eggs.

It is impossible to estimate current numbers but there has been a severe drop over the last 25 years. Their decline is due to demand for the adult's cartilage for soup, their oil for cosmetics and skin for leather. Eggs are taken in thousands from beach nests and the young that hatch are preyed on by birds, fish and crabs. They are now protected in some nesting and feeding areas, such as in Queensland, Australia, where 5,000 km of beach and 2,000 km of the Great Barrier Reef are designated refuges.

IUCN, or the International Union for the Conservation of Nature is the major conservation body now trying to preserve nature in the wild, and natural resources all over the world. It was started in Brussels in

White-bearded gnu and young

Aye aye

Golden lion tamorin

Mountain gorilla

Asiatic lion

Tiger

this destruction because he has altered their environment or killed them in such numbers that they have not been able to survive. Today, one in every 40 species of birds and mammals is in danger of dying out. Many other animal groups such as amphibians, reptiles and fishes also have many species in danger of becoming extinct due to man's pressures.

Man hunted wild animals originally for food and clothing, and later just for sport or to be put in museums or zoos. Other animals have been destroyed because they have been looked upon as vermin, or poisoned by PESTICIDES designed to exterminate weeds or insect pests. The disappearance of many species has been caused by man's destruction of habitats, by felling trees for timber, or by increasing the agricultural or urban land area, or digging up land for building purposes.

Today, two major international conservation bodies are trying to preserve nature in the wild as well as our dwindling natural resources. These are the International Union for the Conservation of Nature and Natural Resources (known as the IUCN) and the WORLD WILDLIFE FUND (WWF). Through their work in raising funds and helping to convince countries to pass and enforce conservation laws, many animals are now protected from being harmed or endangered by man.

The orang-utan, a large ape with a reddish coat, is one of the most endangered species of mammals in south-east Asia (the islands of Sumatra and Borneo). Its numbers have dwindled in the wild mainly due to its tropical rain forests being destroyed to increase agricultural land and because poachers now kill adult females to obtain the live young for the animal trade. However, the numbers now being bred in captivity are encouraging and plans to return zoo-bred specimens to the wild are underway,

although it is very difficult to encourage tame specimens to become independent of man.

The TIGER of China, northern Asia and India is in a serious plight at the moment due to trophy-hunters, and habitat destruction. Although less than 2,000 survive in India, they are now protected. The largest of all tigers, the nomadic Siberian tiger, hunted for the pharmaceutical trade, used to range from Siberia through Mongolia to Korea and Manchuria. Its population is now stable in only one protected mountain system in Korea, a remote corner of Manchuria and some protected areas of the far eastern USSR.

Everything in nature is interdependent. It is impossible to destroy anything – to fell a tree or bulldoze a patch of wasteland, without affecting innumerable creatures and upsetting nature's delicate balance. For example, because of man's interference in killing natural predators, certain

Above: The main causes for the threatened survival of these endangered species are trophy-hunting, forest destruction, the capture of live specimens for the animal trade, killing outright for man's needs or being directly in competition with man and his domesticated animals.

Below: The European bison, or wisent, almost became extinct due to the hostilities of the First World War. From a protected nucleus its numbers were rebuilt to a fairly healthy level. It survives mainly in zoos and the Bialowieza Forest in Poland.

1934. One of its major tasks has been to collect information about species in danger of extinction and begin action to prevent total loss. The IUCN co-operate with the WORLD WILDLIFE FUND and governments to conserve endangered species. For example, the POLAR BEAR is now protected by the countries in which it ranges – the USA, Canada, Norway and the Soviet Union.

J **Japanese crested ibises** are some of the rarest birds in the world. They are confined to Sado

Island and possibly the Noto Peninsula, Japan. Once they were found across northern China, Manchuria, Korea and Japan. They declined to 8 known specimens this century with the deforestation of their favoured wooded wetlands and virgin forests. They are unlikely to survive in the wild.

M **Mauritius kestrels** are small falcons that have declined to drastically low numbers as they were shot for supposedly killing chickens. The species may well be extinct.

Mountain gorillas are the largest and most powerful of all living primates and inhabit the dense remote

Female mountain gorilla

forests of West Africa. Their decline is due to shooting by local tribesmen for food, by animal traders wanting the

young, and also through forests being taken over by agriculturalists. Females only bear young every 3 to 4 years so the reproduction rate is not very high.

N **National Parks** are areas of protected land, found worldwide. The first to be established was the Kruger National Park in South Africa, an area the size of Wales or Belgium, and now still among the finest and most visited national parks.

Nature reserves are areas set up mainly in European

Javan rhino

Blue whale

Green turtle

Mauritius kestrel

Japanese crested ibis

Above: The flightless dodo, once common in Mauritius, became extinct in about 1680. It was hunted by sailors for sport and food, and its nesting sites were disturbed by pigs.

Below: Przewalski's horse is the sole survivor of the wild horse. Species exist in many major zoos throughout the world and they may be re-introduced into the wild later this century.

Below: Due to better political relationships with China, where the giant panda survives in the wild, more pairs of this delightful animal are being displayed in western zoos, having been donated as gifts to visiting dignitaries. Peking Zoo has had great success breeding them in captivity, but the first baby born in the west only occurred in 1977 at Chicago Zoo in the USA.

animals such as rabbits and rats, have increased abnormally in number and do millions of pounds worth of damage to crops every year. The introduction of new animals can also destroy the habitat itself, as in the Galapagos Islands where the voracious goat has destroyed areas of trees and bushes which support other wildlife.

In many areas of the world man has spoilt the basic natural resources such as water, air or soil by polluting them to such a degree that the balance of nature has been upset. For example in

Banff National Park, Canada

countries to protect the wildlife that lives within its area. They range from the size of a small pond to millions of hectares in area.
Né-Né, or Hawaiian goose, was saved from extinction by breeding the 30 survivors in captivity and then later reintroducing some into their native islands. It is now the official bird of the state of Hawaii.

O **Oil pollution** is still a serious threat to marine animals. Birds suffer the most, because as well as being totally clogged with oil

so that they are unable to move and feed, their feathers may be damaged so they are no longer waterproof. This causes the bird to

Né-Né

become waterlogged and drown.
Orang-utans are heavily-built primates with a reddish coat that live in the jungles

Arabian oryx

many cities, such as Tokyo and San Francisco, the pollution by smoke from homes, factories and car exhaust fumes, was so great that the health of people was seriously affected. Some governments have now passed Clean Air Acts so that air pollution is becoming less of a problem, although enforcement is not always easy.

Oceans and freshwater rivers and lakes are still endangered, however, due mainly to factory and home waste and OIL POLLUTION. Oil leakages from tankers or sea oil rigs are a common event these days and the damage to marine life is enormous. The *Amoco Cadiz* in 1978 spilt 250,000 tonnes of oil into the sea off Brittany. The beaches were polluted and chemical dispersants were used to break up the oil. Although every effort is made to 'mop-up' such a disaster, thousands of seabirds and fish perish. The

Above: One way in which the survival of a species can be ensured is by studying its habits in the wild. Here a black rhinoceros female and young are being filmed in the Serengeti.

Above left: The Trans-Amazonian highway crosses about 2,000 km of virgin tropical rain forest in almost a straight line from Belem in northern Brazil to Brazilia. The opening up of these new areas to man's exploitation must endanger the survival of more animal species.

problems created by polluted fresh water range from outbreaks of disease in animals to the destruction of the habitat and the disappearance of fish and other wildlife in the area. Facing the dangers of contamination, the United States, Great Britain and West Germany have taken the lead in constructing long-range programmes to try to avoid future catastrophies and counteract present polluted areas.

Whales, the mighty monarchs of the oceans, have not disappeared due to the polluted sea but due to overhunting by man. Some species have been reduced to such low numbers that although now protected, they may not survive as they cannot reproduce quickly enough to ensure that their numbers increase. The 20th century has produced the greatest slaughter of whales. It was intensified at the beginning of the century to such

of Sumatra and Borneo. People cannot now keep them as pets; they are sent to centres where they are retrained for life in the wild.
Oryx is the name given to 3 species of rare gazelles. The Arabian oryx is the smallest and rarest, its decline being mainly due to the discovery and extraction of oil from its habitat. The other 2 are the scimitar-horned and the beisa oryx (more common).
Ospreys, or fish hawks, are examples of protected birds of prey. They were formerly killed because it was thought that they took domestic ani-

mals. Now, 24-hour watches are kept on nests in the breeding season.

Wolf

P **Passenger pigeons** thrived in their millions in their native North America until the late 19th century when they were shot and ruthlessly slaughtered. The last died in captivity in 1914.
Pesticides poison many land animals who may eat them unknowingly. Often the chemical DDT is absorbed and builds up in the animal's body.
Pollution is usually brought about as a by-product of man's activities. The air is polluted by smoke such as sulphur dioxide and waste gases. Water is damaged by

Polar bears

Below: The culling of seals is an emotive subject at the moment. A certain amount of culling may be beneficial in lowering the infant mortality rate in rookeries due to overcrowding. However, the cubs are clubbed to death to avoid damaging the seal skin. Many people believe this method is cruel and should be stopped.

Right: Although many whales such as the blue, right, humpback and grey are totally protected, their numbers are so very low that they might not be large enough to ensure their survival. Closed hunting seasons and size of catch limits protect the other species to some extent, but numbers still fall.

Right whale
10,000
2,000

Fin whale

1946   1976

Blue whale
210,000
450,000
13,000
100,000

Sperm whale
1,000,000
620,000

Humpback whale
100,000
7,000

Grey whale
15,000
11,000

an extent that the North Atlantic right whale (*Eubalaena glacialis*) was virtually exterminated. A hundred years ago, a whaler's three-year trip netted him 37 whales. Today, with modern detecting equipment, weapons and fast boats, a whale a day (even three or four) is the usual catch. At the present time, whaling is theoretically a regulated industry with each country being allowed a catch of a specified number of whales a year. In principle, all species are protected by international agreements arrived at by the International Whaling Commission. However, the USSR and Japan are not conforming to the agreed terms, despite pressures from conservation bodies and governments of other nations.

GAME RESERVES, NATURE RESERVES and NATIONAL PARKS are found all over the world today so that animals and plants can exist within a protected area in their natural habitat. There are, of course, still problems from poachers and from thoughtless people who may start fires, or disturb breeding animals by going too close.

Throughout the world, man has put at risk areas from a few square metres to thousands of millions of hectares. In Britain millions of kilometres of hedgerows have been destroyed to create large-scale fields so that modern agricultural machinery can function. In the Amazon basin in South America the forest is being cut down and long roads penetrate the dense jungle which was hitherto considered to be unassailable. Here the whole forest life is at risk. However, some local, national and international organizations are beginning to see that this work must be curbed before the balance of nature is destroyed forever.

Above: Oil pollution caused by disasters, such as the running aground of the Amoco Cadiz off northern France in 1978, causes the death of millions of birds, as well as marine creatures.

poisoning by industrial effluents, or by deoxygenation due to the decomposition of organic matter such as sewage.
Polar bears are dwellers of the Arctic tundra and icebound north pole region. Their numbers were severely reduced due to game hunting. Recently they were being shot from helicopters. Then their habitat was interferred with by the opening up of the northern polar region for oil. Polar bears are now protected by the countries in which they are found and numbers are increasing.

Przewalski's horse is the only wild horse to survive in the wild, mainly in Outer Mongolia. Numbers in captivity now ensure its survival. Its decline in numbers was due to hunting for meat and being bred with domestic ponies.

Q **Quaggas** were a common zebra species which used to graze the grasslands of the veldt of South Africa. When colonizers moved in, this mammal with its striped neck and head was killed indiscriminately because it competed with the cattle for food. The last specimen died in captivity in 1883.

S **Siamese crocodiles** survive only on a crocodile farm in Thailand. Most crocodile species are endangered because their skins make excellent leather.
Spanish Imperial eagles are distinguished from other large eagles by their white shoulders. Not known to breed outside Spain (where total numbers are less than 100), they are protected only in Coto Donaña National Park. Seven pairs have nested there in the cork oaks. The species will not survive without protection.

Booted eagle

T **Tasmanian wolves,** or thylacines lived in captivity until the 1930s although none were bred. Their decline was due to settlers shooting them as they preyed on sheep flocks. Although occasional sightings have been recorded in the western part of Tasmania, they are probably extinct.
Tigers are depleted in number over most of India, northern Asia and China. Probably less than 2,000 survive today. The decline is due to trophy-hunting, poaching for the fine skin,

Below: Oil pollution from supertankers that have collided or broken up at sea are almost weekly happenings around the world. The crude oil sticks to the birds' feathers and bills thus preventing them from flying and fishing.

Below: Rivers all over the world have become polluted to such a degree that most of the aquatic life including fishes, plants and insects are endangered. The main cause of the pollution is the release of highly toxic industrial wastes and town sewage.

Below: Rubbish heaps have thousands of tonnes of waste poured on them every day. Much of it is leftover food and this attracts rats as well as flies and birds. Fleas living on rats transmit diseases such as the plague and typhus.

Below: Open-cast mining, deforestation and agricultural programmes have destroyed the natural habitat of many wild animals, causing their numbers to be reduced often to the point of endangering the species.

Below: Air pollution threatens not only birds and insects but also man. Some of the intermediate chemicals in the complex process of producing plastics and similar oil derivatives are frighteningly toxic if released into the air.

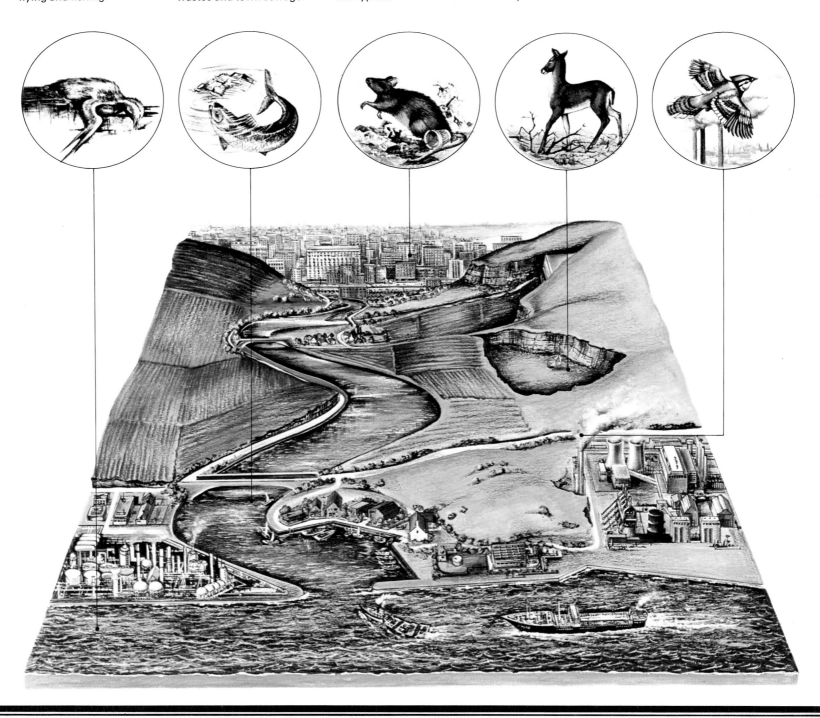

destruction of the natural habitat and because of their supposedly man-eating and cattle-killing habits. They are now legally protected over most of their range, and some reserves have been established. However, they will quite probably be extinct in the wild by the year 2000.

Trumpeter swans are the largest North American swans. At one time numbers were down to below 60 in the states excluding Alaska, but now total over 2,000. Hunting was the major cause of decline. Un-

Gila monster, a poisonous lizard

doubtedly this bird was saved by the creation of the Red Rock Lake refuge in Yellowstone Park, Wyoming, USA, in 1935.

W **Whooping cranes** are rare North American birds that breed on remote sub-Arctic lakes and migrate 3,700 km to winter in Texas.

Unfortunately they risk being shot by sportsmen on the way in mistake for sandhill cranes.

World Wildlife Fund, or WWF, was started in the 1960s when it became obvious that something had to be done to halt the growing threat of extinction facing large numbers of plants and animals, and also to reduce the rate at which natural habitats were being destroyed by man. Its main task is to raise money to finance conservation action. The headquarters are in Switzerland with branches worldwide.

World Wildlife Fund emblem

The Plant World

Mark Lambert

Introduction

The Plant World is a detailed account of the plant kingdom, ranging from single-celled algae to the giant redwoods of California, one of which is more than 111 metres high. It includes a description of how plants have evolved throughout Earth history and there is an important section on plant biology, including descriptions of their cellular structure and how plants make their food, grow and reproduce. Like animals, plants have adapted to a wide variety of environments and **The Plant World** tells of the extraordinary adaptations which have enabled plants to colonize most parts of the globe. Without plants, our Earth would be lifeless. They provide us with food and drink and many other items, ranging from timber and pulp for paper to chemicals used to make medicines, poisons and drugs. Yet many plant species are in danger of extinction, because of the activities of man.

No life would be possible on Earth without plants. From the study of fossils, we know that single-celled algae and bacteria were among the first living things on Earth. Plants have since adapted successfully to almost every known habitat.

Plants Great and Small

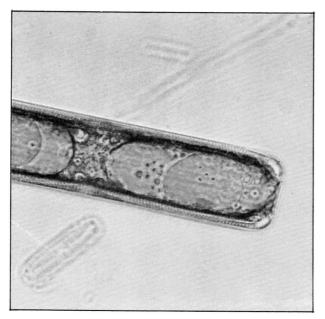

Left: A photomicrograph of the diatom *Pinnularia*. Diatoms belong to the group called algae and are the smallest of all plants.
Below: If a finger nail was enlarged 10 times, a diatom on the same scale would still only be a dot 0.25 mm in diameter.

Below right: A Californian coast redwood (*Sequoia sempervivens*) is one of the tallest plants in the world. It may grow to a height of 60-85 metres, and the largest specimen is over 111 metres tall. The redwood is a conifer that may live to be 1,800 years old.

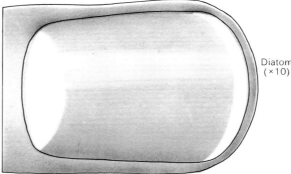

Diatom (×10)

Plants grow in nearly all parts of the world and are found in an incredible variety of shapes and sizes. There are over 360,000 known species of plants in the world. They range from single-celled algae that can only be seen under a microscope to giant redwood trees over 100 metres tall. Some plants live for only a day, others for thousands of years. Other plants are so rare that they only grow in one small area of the world, and even there they prove hard to find. Many plants on the other hand are very common, particularly the grasses that cover a large part of the land.

What are plants?

Many plant varieties are grown for their beauty or for food. However, in addition to man's needs, plants are essential for all life on Earth. Only plants, with the help of sunlight, can build up living material from water, minerals and air.

Like animals, plants have the particular characteristics of living organisms. These are feeding, RESPIRATION, EXCRETION, growth, movement, sensitivity, and REPRODUCTION. However, plants differ from animals in the way in which they perform some of these functions. Animals obtain their food by eating plants or other animals. Plants on the other hand, make their own food by PHOTOSYNTHESIS. Most animals can move from one place to another, and many have muscles that they use for movement. The nervous systems of animals are also used for movement, and for sensing changes in the environment. Plants do not have muscles or nervous systems, and most plants remain in one place. But certain movements, such as phototropism (*see page 111*) do occur.

Of course, there are exceptions to these rules. Fungi could be called the renegades of the plant world as they cannot make their own food by photosynthesis. Some small algae have whip-like organs that enable them to swim.

Reference

A **Algae** are the simplest groups of plants, belonging to the division Thallophyta. Algae range from single-celled plants to giant seaweeds.
Angiosperms, or flowering plants, are the most advanced group belonging to the division Spermatophyta, subdivision Angiospermae (*see pages 100–112*).

B **Bennettitales** are an extinct order of gymno-

sperms (seed-bearing plants) that existed from the Triassic to the Cretaceous periods.
Biomes are major ecologi-

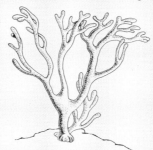

Codium fragile, an alga

cal regions of plant and animal life. The plants of a biome make up a FORMATION.
Botany is the study of plants.
Bryophytes are plants belonging to the division Bryophyta. They are divided into 2 classes — the liverworts (Hepaticae) and mosses (Musci) (*see pages 90-93*).

C **Cambrian period** (570-530 million years ago). The period gets its name from *Cambria*, the Latin name for Wales.
Carboniferous period

(345-280 million years ago). The period was noted for its vast swampy forests and the appearance of the first seed plants.
Class is one of the groupings used in the classification of plants and animals. A class is divided into ORDERS. In the plant kingdom several classes form a SUBDIVISION.
Classification is the way in which plants and animals are divided into groups and sub-groups. The largest group is a KINGDOM, and the smallest group is a SPECIES.
Climax, see SUCCESSION.
Conifers are a group of

Bennettites

cone-producing, woody plants, belonging to the gymnosperms. They make up the order Coniferales.
Cordaitales are an extinct order of gymnosperms that

The evolution of plants

The Earth was formed about 4,800 million years ago. As it cooled, the first tiny organisms probably formed in the warm seas. The earliest life forms for which we have evidence are blue-green algae. These plants produced lime secretions that later hardened. As a result we now find fossils called stromatolites which consist of many layers of lime laid down by these blue-green algae. Stromatolites dating from 3,100 million years ago have been found in Rhodesia.

The next advance in plant evolution took place in the SILURIAN period. During this time a group of plants called PSILOPHYTES existed. These were marsh plants that probably grew around the edges of lakes. Very few fossils of these plants have been found, but we know that they had tall, branching stems without leaves. At the end of each branch there were one or more capsules that contained spores. Psilophytes were probably the ancestors of the two modern psilotes (primitive

plants), *Psilotum* and *Tmesipteris*. No intermediate fossils have been found, and the psilophytes appear to have died out by the end of the Devonian period.

Before the Devonian period we know very little about how plants evolved. But during the Devonian, Carboniferous and Permian periods there existed vast humid forests of plants. This mass of vegetation eventually died and formed peat which compressed into the present-day coal seams. There were several groups in existence at that time: FERNS, horsetails, LYCOPHYTES, PTERIDOSPERMS and CORDAITALES.

The earliest ferns had thick stems and small leaves, but later ferns were very similar to those of today. The lycophytes were the ancestors of modern clubmosses and quillworts. Those that existed during the Carboniferous period, such as *Lepidodendron* and *Calamites*, were tree-sized plants. They produced their spores in the cones like modern members of this group which have

Above: The Patriarch Tree in the Sierra Nevada, USA, is the world's largest bristlecone pine (*Pinus aristata*). Some bristlecone pines are the world's oldest known trees. The oldest recorded was 4,900 years old. It grew on the north-east face of Wheeler Peak in California, USA. The oldest-known living tree is the bristlecone pine named Methuselah, which is 4,600 years old and grows in the Californian White Mountains.

---

existed from the Devonian to the Permian periods.
Cretaceous period (135-65 million years ago). *Creta* is the Latin for chalk.
Cycads are a group of gymnosperms, belonging to the order Cycadales. They were more widespread during the Jurassic and Cretaceous periods, but 9 genera still exist today.

D de Candolle, Augustin (1778-1841) was a Swiss botanist who was the first to classify plants by their similarities and differences.
Devonian period (410-345

Cordiates

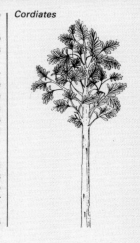

million years ago). Primitive ferns and lycophytes were evolving at this time.
Division is one of the groupings used in the classification of plants. A division is a sub-group of a KINGDOM, and is divided into several SUBDIVISIONS.

E Ecology is the study of plants and animals in relation to their surroundings and to each other.
Environment. This term includes all the conditions in which a plant or animal lives, such as temperature, light, water, and other plants

and animals.
Excretion is the process by which a plant or animal gets rid of its waste products.

Stone pine (a conifer)

F Family is one of the groupings used in the classification of plants and animals. Several families make up an ORDER, and a family is divided into genera (see GENUS).
Ferns are a group of spore-producing plants related to the LYCOPHYTES. They belong to the division Pteridophyta, subdivision Filicophyta.
Formations are the main natural types of vegetation of the world. They are communities of plants extending over large areas. The type of plant life within a formation is determined by the climate.

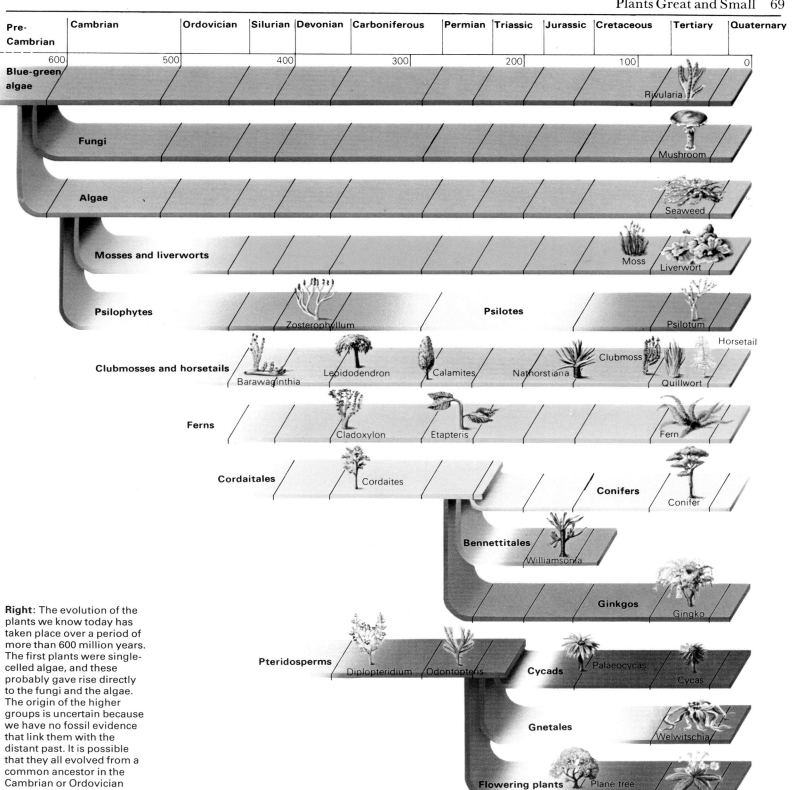

Pre-Cambrian	Cambrian	Ordovician	Silurian	Devonian	Carboniferous	Permian	Triassic	Jurassic	Cretaceous	Tertiary	Quaternary
600	500	400			300		200		100		0

Right: The evolution of the plants we know today has taken place over a period of more than 600 million years. The first plants were single-celled algae, and these probably gave rise directly to the fungi and the algae. The origin of the higher groups is uncertain because we have no fossil evidence that link them with the distant past. It is possible that they all evolved from a common ancestor in the Cambrian or Ordovician periods.

cones at the top of the shoots.

At the same time as these spore-producing plants existed, the Cordaitales and pteridosperms had developed the ability to produce seeds. Seeds are a more efficient method of reproduction than spores because each plant starts life as a many-celled embryo that is well protected inside the seed case. As a result, seeds survive more easily than spores.

The Cordaitales were tall trees that formed large forests in the Carboniferous period. But by the end of the Permian period they were extinct. The conifers, Ginkgos and BENNETTITALES evolved from the Cordaitales. The Ginkgos flourished during the Cretaceous period, but today the only survivor of this group is the Maidenhair tree, *Ginkgo biloba*. The Bennettitales looked like CYCADS and bore seeds in structures that resembled flowers. But despite this, they were probably not related to either cycads or flowering plants.

The pteridosperms, or 'seed ferns', resembled true FERNS in many ways except that they produced seeds at the ends of special branches. This group flourished during the Carboniferous period, but became extinct during the Triassic period. The cycads, Gnetales, and the flowering plants evolved from the pteridosperms.

Above: The monkey puzzle tree (*Auracaria araucana*) is one of the few trees that live at high altitudes. It is also one of the few conifers found in the Southern Hemisphere, living in the Andes Mountains of South America.

Cycads are palm-like plants. Many types existed during the Cretaceous period, but only nine genera exist today. The Gnetales are an odd group of only three genera — *Welwitschia, Ephedra,* and *Gnetum.* They are classed as gymnosperms but show many characteristics of flowering plants.

The most important group, however, is the flowering plants, or angiosperms. During the Cretaceous and Tertiary periods they increased rapidly in number, and today they are the dominant group of plants. There are two main reasons for their success. First, they have been able to adapt to almost every habitat in the world. Secondly, they have extremely efficient methods of producing and dispersing seeds (*see page 104*).

Where plants live

The type of vegetation that can be found in a particular area is determined by the ENVIRONMENT. Both the climate and the geography of the region affect the nature of the plants that are found. The natural world can be divided into several BIOMES, each of which contains a distinct plant FORMATION. The richest of all the biomes is the tropical rain forest. Vegetation of this type occurs in areas near the equator that have a high

Below and right: The alpine areas of the world contain many unique plant forms. In the highest regions only lichens and mosses can survive the cold. But well below the snow-line coniferous forest is mixed with rich pasture.

Below and left: The deciduous forests of the world contain broad-leaved trees such as the beech and silver birch. Many other plants, including ferns and mosses can be found growing in the leaf litter among the trees.

rainfall. A tropical climate is suitable for continuous plant growth, and its rain forests contain large trees, usually evergreen, together with other exotic species of plants.

In climates where growth cannot be continuous because of seasonal variations in temperature, the type of vegetation changes dramatically. In temperate regions two main biomes occur — grassland and temperate forest. In grassland regions, herb vegetation dominates, together with some shrubs and trees. Temperate forests may be deciduous or coniferous. The former shed their leaves in autumn whilst the latter are evergreen. The exact nature of the plant life depends very much on the seasonal variations of climate. For example, if there is any winter frost, many plant species cannot survive. Many non-woody plants have adapted to these conditions by remaining dormant during the winter months.

The harshest biomes are those where it is either very cold or very hot. In such areas the density of plant life is very much reduced. Montane biomes include the tundra and mountain regions, where it is so cold that only the hardiest plants can survive. This is also true of desert conditions, and there are many other environments where only specially adapted plants can live (*see pages 113–120*).

Above: Life began in the oceans of the world and they still contain a wide variety of plant life, such as *Laminaria,* a brown seaweed. *Laminaria* belongs to the algae group which includes most of the sea plants.

The study of plant formations and world vegetation zones is included in the science of plant geography. This subject also deals with the spread of plant groups and species. For example, we know that cacti originated in South America, but birds and man have carried their seeds far afield. They are still gradually spreading into all the areas to which they are suited.

Plant geography also includes the study of the effects of natural barriers, such as mountains and oceans. For example, *Pinus sylvestris,* the Scots pine, is one of the dominant trees in northern Europe. However, it is not found in the natural vegetation of North America. The Atlantic and Pacific Oceans have prevented this species from spreading.

PLANT ECOLOGY is concerned with the more detailed study of the vegetation of an area or locality. For example, the ecology of an area of deciduous woodland or chalk grassland can be studied to learn which species of plants are present. All the plants living in such habitats are suited to the local conditions of climate and soil. At the same time, they are suited to living together within a plant community. Plants are influenced to a great extent by others around them, particularly when there is competition for sunlight, space and soil nutrients.

Below and right: The tropical regions of the world include many areas of rain forest. In this forest lianas – the long stems of woody climbing plants – can be seen together with a climbing palm tree.

Below and left: The map of the world shows that the tundra includes part of Alaska. Few plants can grow in the Arctic regions of the northern tundra, but Alaska can support some hardy flowers in summer.

of spore-producing plants, belonging to the division Pteridophyta. This includes all the ferns, lycophytes, psilophytes and psilotes.
Pteridosperms are a group of extinct gymnosperms that existed from the Devonian to the Triassic periods. They belong to the order Pteridospermales.

Q **Quaternary period** (2 million years ago to the present day). The name comes from the fact that geologists formerly divided fossil-bearing rocks into 4 periods.

R **Reproduction** is the process in which an organism produces a new individual.
Respiration is the process in which an organism takes in oxygen and uses it to 'burn' food to provide itself with energy. During the process carbon dioxide is formed and released (*see pages 72-80*).

S **Silurian period** (440-410 million years ago). The name comes from the *Silures,* an ancient tribe who lived in the area of Wales where rocks of this period

were first studied. The first land plants were probably evolving at this time.
Species is the smallest group used in the classifica-

Pteridosperm

tion of plants and animals. Members of the same species can successfully breed with each other but they cannot breed with members of another species.
Subdivision is one of the groupings used in the classification of plants. A subdivision is divided into CLASSES, and several subdivisions make up a DIVISION.
Succession is the gradual, progressive change from simple, hardy plants, such as lichens and mosses, towards more advanced plants during the colonization of a

piece of land. When the succession is complete, and the plant community is stable, a climax has been reached.

T **Taxonomy** is the detailed study of the classification of plants and animals.
Tertiary period (65-2 million years ago). The name is derived in the same way as the QUATERNARY PERIOD.
Triassic period (225-195 million years ago) is so named because of a 3 layered area of rock in Germany that dates from this time.

To understand the intricate mechanisms of all living things, including plants, we must study the minute cells of which they are composed. Each cell works like a microscopic factory, making its own special contribution to the miracle of life.

How Plants Work

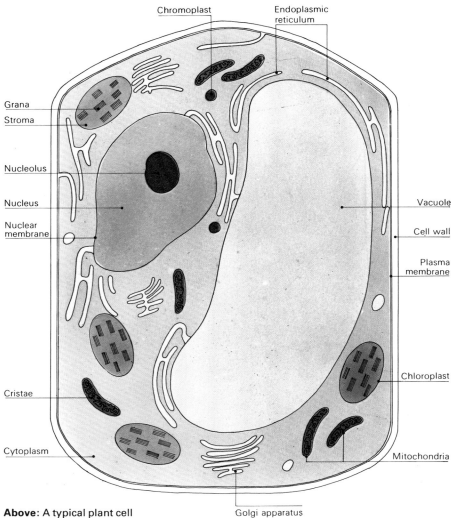

Chromoplast

Endoplasmic reticulum

Grana

Stroma

Nucleolus

Nucleus

Nuclear membrane

Cristae

Cytoplasm

Vacuole

Cell wall

Plasma membrane

Chloroplast

Mitochondria

Golgi apparatus

Above: A typical plant cell consists of a cellulose cell wall that encloses the cytoplasm and the nucleus. In the cytoplasm there are several structures that do the work of the cell. The nucleus controls and co-ordinates this work. The vacuole is a space inside the cell filled with a watery fluid.

Plants and animals are made up entirely of CELLS. A plant cell is like a tiny enclosed box. It contains all the chemicals and structures that the plant needs in order to live. Neighbouring cells work together so that the functions of the whole plant are co-ordinated. But at the same time, each cell works by itself like a tiny factory.

The outer covering of a cell is called the cell wall. Each cell is connected to its neighbour by tiny holes in these cell walls. Chemicals can pass through these holes from one cell to the next. The contents of each cell consist of the NUCLEUS and the CYTOPLASM.

The nucleus is the central controller of the cell – like a computer centre in a factory. It contains thread-like structures called CHROMOSOMES. These contain all the information needed to organize the activities of the rest of the cell. The information is contained in a chemical compound called DNA, and the information is carried to where it is needed by RNA.

The cytoplasm consists of a watery mixture that includes a number of important solid structures called ORGANELLES. These are the working parts of the factory. Some organelles use the chemicals imported into the cell to make other chemicals that are useful to the plant, such as SUGARS and PROTEINS. Other organelles help to produce energy for the plant.

Proteins are a very important group of chemical products. They are made on a thin folded membrane in the cytoplasm, called the ENDOPLASMIC RETICULUM. The proteins that a cell makes may be used for growth or for repairing damage. On the other hand, they may be special proteins called ENZYMES. These are essential for the plant. Without them, many of the chemical reactions that take place in cells could not occur.

All the activities of a cell require energy. This is provided by the MITOCHONDRIA, which are the

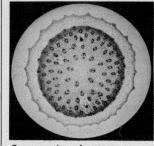

'power-houses' of the cell — like electrical generators in a factory. Respiration is the energy-producing process and it takes place inside the mitochondria.

Like full-scale factories, cells have to get rid of their waste products and export some of the materials they make. The main waste product is the carbon dioxide produced during respiration. This diffuses through the cell wall. The remaining waste products either diffuse out, or they are stored in the cell in a solid, insoluble form. Many chemicals exported by the cell also diffuse out, but molecules of some elements are too large to pass through the cell membrane. They are probably removed from the cell by a structure called the GOLGI APPARATUS.

Cells for different purposes

The simplest cells of all are found in the regions where growth is taking place. They are small, cube-shaped cells with thin cell walls, and completely filled with cytoplasm. As the region of

Left: This photograph was taken through an electron microscope. It is a portion of a plant cell wall – magnified 30,000 times. The cellulose fibres of the wall can be seen clearly.

Above: Cork comes from the bark of the cork oak (*Quercus suber*). Robert Hooke (1635-1703) discovered that cork is made up of millions of tiny compartments, which he called cells.

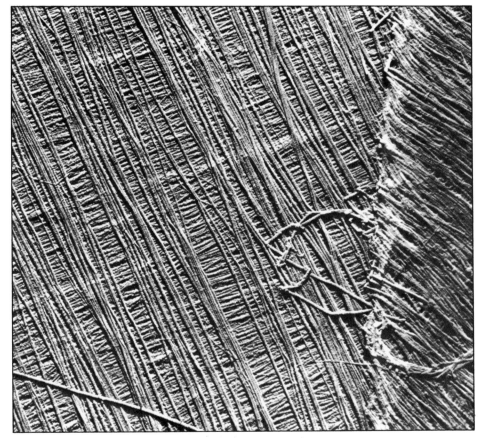

growth moves away, they begin to grow. However, the amount of cytoplasm increases very little. Thus, as a cell grows larger, its cytoplasm stretches and a space, or VACUOLE, is formed.

The eventual function of a cell depends on its position in the plant. Therefore, as cells grow they become differentiated — that is, cells with different functions differ in shape and structure.

A large proportion of cells become PARENCHYMA CELLS. These make up a great deal of the plant body, especially in young plants. They have thin cells walls made up of a substance called CELLULOSE. They often contain storage products, such as starch grains. COLLENCHYMA CELLS are firmer cells. They have cell walls that are thickened with cellulose, often in the corners of the cells. The stems of the larger non-woody plants need the firm support given by collenchyma cells. Where even more support is needed, non-living stiffening cells are found, such as SCLERENCHYMA FIBRES and STONE CELLS. These

In older, thickened cells, a secondary wall is laid down inside the primary wall. This may be made of cellulose, or lignin and cellulose, or lignin and cutin. It is generally perforated by pits, which allow connections between cells. Between the cell walls of adjacent cells lies the MIDDLE LAMELLA.

Chloroplasts are small chlorophyll-containing bodies in the cytoplasm of a cell. This is the site of photosynthesis. A chloroplast consists of a double membrane surrounding a material called the stroma. Embedded in the stroma are grana, and it is here that photosynthesis takes place. The grana are composed of a number of round disks, which are stacked like coins. The disks carry the chlorophyll.

Chlorosis is a disease of plants in which the leaves turn yellow. This is because the plant cannot make chlorophyll, either through lack of light or lack of certain minerals, such as magnesium.

Chromatin is the granular material in the nucleus of a cell. It consists of PROTEIN, DNA, and RNA.

Chromoplasts are small bodies in the cytoplasm that contain pigments. Those that contain chlorophyll are called CHLOROPLASTS. Other chromoplasts may contain carotenes, xanthophylls, anthocyanins or other pigments.

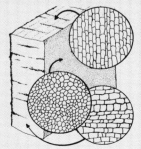

Cork cells

May blossom

Chromosomes are thread-like bodies in the nucleus of a cell. They consist mostly of DNA and protein. A normal cell has several pairs of chromosomes. The 2 members of each pair are identical in appearance. They are called homologous chromosomes. The body cells of a particular species always contain the same number of pairs of chromosomes. The broad bean (*Vicia faba*) has 6 pairs; the onion (*Allium cepa*) has 8 pairs. Man has 23 pairs.

Collenchyma is a cell tissue that helps to provide support

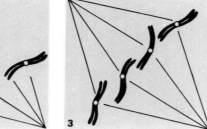
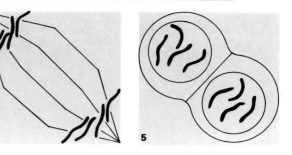

Above: In mitosis 2 new cells are formed. **1.** At the beginning of division the chromosomes are already divided into chromatids. The nuclear membrane is still present. **2.** A fibre-like structure, the spindle, forms. The nuclear membrane disappears. **3.** The chromosomes are arranged on the centre of the spindle. **4.** The chromatids move apart and move towards opposite sides of the cell. **5.** A new cell wall forms down the middle of the cell. Two new nuclear membranes form around the new chromosomes. The number of chromosomes in the 2 new cells is the same as the number in the original cell.

Left: A plant in full turgor is erect and healthy.

Water and food move from cell to cell relatively slowly, and so plants have developed 'plumbing systems' that can move materials more rapidly through the plant. XYLEM cells conduct water up the plant from the root to the leaves. They are non-living, elongated cells that are joined end-to-end to form a continuous system of pipes. The walls of xylem cells are thick. A secondary wall is present, made up of cellulose and lignin. In some cases it may take the form of rings around the inside of the xylem cell. Even in heavily-thickened xylem cells the secondary wall is perforated by pits.

PHLOEM cells conduct food and other material up and down the plant. Some are elongated like xylem cells, but they are not open at both ends. Instead, the end walls that lead from one cell to another are perforated like sieves. Hence, the dividing wall between these phloem cells is called a sieve plate, and the cell is known as a sieve tube. It contains cytoplasm, but it does not have a nucleus. As a result, each sieve tube has to be controlled by one or two companion cells.

Dividing cells

There are two processes by which cells divide — MITOSIS and MEIOSIS. As a plant grows larger, new cells are needed. These are produced by mitosis, which occurs in all the growing parts of a plant. Meiosis produces sex cells and occurs in the reproductive organs of a plant.

The aim of mitosis is to produce two new cells that are identical in every respect to the original cell. Most parts of the cell divide quite simply, but the nucleus has to undergo a more complicated process in order to make sure that the new cells have the same number and type of chromosomes.

During the course of mitosis, each chromosome divides into two chromatids. The chromatids are drawn apart, one to each end of

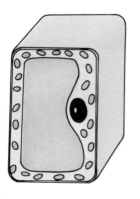

Above: The cells of a plant in full turgor contain the maximum amount of water possible. Hence they are stiff and strong.

cells have walls that are thickened with substances called CUTIN and LIGNIN.

As well as support, a plant needs protection. Leaves and stems of non-woody plants are covered in a layer of cells called the EPIDERMIS. The exact shape of these cells varies from plant to plant, but they are roughly cube-shaped. On the outer surface of the epidermis is a layer of cutin called the cuticle. Some plants have hairs or prickles, which are composed of modified epidermal cells.

As a plant gets older and becomes woody, the epidermis on its stem and branches is replaced by cork cells, which make up the bark. The cell walls of cork cells are lined with a completely waterproof substance called suberin. As a result, the contents of the cells die, and cork becomes a non-living tissue.

---

for young growing plants. It is found in the stems, leaves and leaf stalks. The cells are usually thickened in the corners, but they are still capable of growing lengthwise.
Cutin is a waterproof substance made up of a number of complicated long-chain molecules. The basic chemicals that form these long chains are carbon compounds called fatty acids.
Cytoplasm consists of all the contents in a cell except the nucleus.

D **DNA** (Deoxyribonucleic acid) is the chemical

compound in the nucleus of a cell that carries the genetic information. It is made up of long strands of linked sugar and phosphate molecules. Attached to the sugar groups are 4 chemical compounds called bases — adenine, thymine, guanine, and cytosine. In a complete molecule of DNA, 2 strands coil round each other in a structure called a double helix that resembles a spiral staircase. The rungs of this 'staircase' are formed by pairs of the bases. Adenine pairs with thymine, and guanine pairs with cytosine.

Silver fir

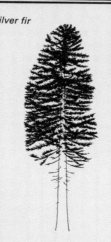

The order in which the bases occur determines the information that is passed from the nucleus by RNA. It also determines the genetic information passed to the next generation.

E **Endoplasmic reticulum.** This is a complicated system of channels present in the CYTOPLASM of a cell. On the surface of its membranes are the RIBOSOMES, and the whole structure is concerned with making PROTEINS.
Enzymes are carbon compounds that act as catalysts

for the chemical reactions that occur in living cells. Thus, they help certain reactions to take place, but they do not actually take part. The enzymes are not changed during the process. Sometimes an enzyme needs to become attached to a coenzyme in order to work.
Epidermis. This is the outer layer of cells of a plant stem, leaf or root. The epidermis of a stem or leaf is usually covered with a waterproof cuticle made of CUTIN.

F **Fats and oils** are carbon compounds that are fre-

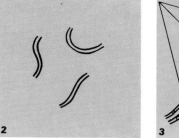

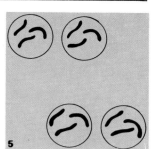
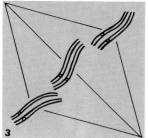

the cell, where they become new chromosomes. When this is complete, the remainder of the cell divides.

Sex cells, on the other hand, are not identical to their parent cell. During reproduction two sex cells fuse together. However, when this happens the number of chromosomes is not doubled. This is because during meiosis the number of chromosomes is halved. Thus, when the sex cells fuse, the resulting cell contains the normal number of chromosomes. Meiosis is therefore often called 'reduction division'. The halving of the number of chromosomes is achieved by two divisions of the nucleus, in which the chromosomes divide only once.

Plants and water

Water is essential for the survival of a plant. A tree consists of about 50 per cent water, and a non-woody plant consists of about 75 per cent water. A plant must have water in its cells so that all the organelles and enzymes can function properly. Without water, the chemical reactions of respiration and photosynthesis could not occur.

Water is also needed for support, particularly in non-woody plants. It is taken up through the tiny ROOT HAIRS by a process called OSMOSIS. The same process causes water to pass from cell to cell. As a cell fills with water, its walls become stretched. When the cell can hold no more water it is described as having maximum TURGOR PRESSURE. In this condition the cell is stiff, and it is therefore turgor pressure that keeps a non-woody plant erect. When there is not enough water in the cells, the plant wilts.

The root hairs of a plant provide an enormous surface area through which water can be absorbed. The water passes through the cells of the root, and from there it is pushed into the xylem vessels of the root by ROOT PRESSURE. The

Above: In meiosis 4 new cells are formed. **1.** The chromosomes appear as long, thin threads. They are *not* split, as in mitosis. **2.** The chromosomes shorten and pair off. **3.** Each chromosome divides into 2 chromatids. Then, still in their pairs, they are organized on the spindle. **4.** The chromosomes are drawn apart. **5.** The 2 cells that have been formed by the first division divide again. But this time the chromosomes do not divide. The chromatids formed in the first division are drawn apart – as in mitosis. Thus the number of chromosomes in each of the 4 new cells is half the number in the original cell.

Right: A plant that has insufficient water wilts.

xylem vessels carry the water up the stem to the leaves. From here a considerable amount of water is lost by evaporation, passing out through the leaf STOMATA. This water loss is called transpiration.

Transpiration has two main uses. Firstly, the evaporation from the leaves has a cooling effect. Plants lose much more water by transpiration on hot days. The rate of transpiration can be controlled by opening and closing the leaf stomata. Secondly, essential minerals are taken up from the soil and carried into the plant by the transpiration stream — the name given to the flow of water from the roots up through the stem and out through the leaves.

At one time botanists could not understand how water moved up a plant. Root pressure alone was not enough to drive water up a tree

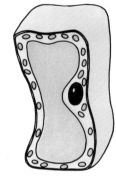

Above: The cells of a wilting plant are weak and bendable. This is because they do not contain enough water to keep the cell wall stiff.

---

quently found as storage products in plants. For example, they are found as food reserves in the seeds of the coconut palm (*Cocos nucifera*) and the cocoa plant (*Theobroma cacao*). They are all derived from compounds called fatty acids.

G Golgi apparatus. This is a small body in the cytoplasm of a cell. It is similar to the ENDOPLASMIC RETICULUM, and consists of membranes folded into sacs, or vesicles. Its function is probably the export of substances out of the cell.

L Lignin is a complicated chemical compound that gives strength to cell walls. It is composed of long chains of carbon compounds.

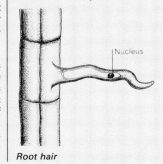

Root hair

M Meiosis is the process of cell division that forms sex cells. During this process the number of chromosomes in the nucleus is halved.

Mesophyll cells are irregular cells that form the spongy tissue in the lower part of the leaf of a flowering plant. There are air spaces between the cells to allow carbon dioxide and oxygen to pass into and out of the cells.

Metabolism includes all the chemical processes of an organism, involving ANABOLISM and CATABOLISM.

Middle lamella. This is the thin layer that cements adjacent cell walls together. It is made of a substance called pectin, together with other similar CARBOHYDRATE compounds.

Mineral salts are chemicals present in the soil, such as potassium nitrate and magnesium sulphate. In the soil they dissolve in water and form ions. In this form they are taken up in the transpiration stream of a plant, supplying it with potassium, magnesium, nitrogen and sulphur.

Mitochondria are micros-

False acacia

copic bodies present in the cytoplasm of a cell. They are the energy-producing ORGANELLES of the cell, and it is here that respiration takes

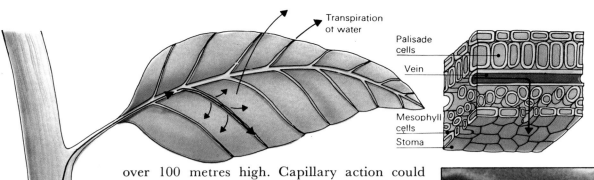

Transpiration of water

Palisade cells

Vein

Mesophyll cells

Stoma

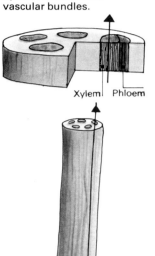

Above and below: Water passes up the stem via the xylem vessels in the vascular bundles.

Xylem Phloem

Water carried up xylem of stem

over 100 metres high. Capillary action could draw up water only a few metres. Various theories were put forward, including some that suggested that there was a pumping action by some of the living cells in the stem.

In fact, water is quite simply pulled up the plant by transpiration. This is made possible only by a very special property of water. Its molecules have a strong tendency to cling together. This property is called cohesion. As water molecules leave the xylem vessels in the leaf, others are drawn in to take their place. Due to this cohesion, the leaf of a 100-metre tree can exert a pull of over 100 atmospheres (103 kilogrammes per square centimetre) — enough to draw water up from its roots.

Plants and energy

The living world obtains all its energy from the Sun. The small fraction of the Sun's energy that reaches the Earth arrives in the form of heat and light. The heat keeps the atmosphere and the surface of the Earth warm enough to support life. Some of the light that falls on the Earth's surface is used by plants to start the energy chain that keeps the living world functioning.

The method whereby plants are able to trap and use light energy is called photosynthesis. Plants use light to make food. The food may be used for the plant's own energy needs, or the plants may be eaten by animals. In this way the

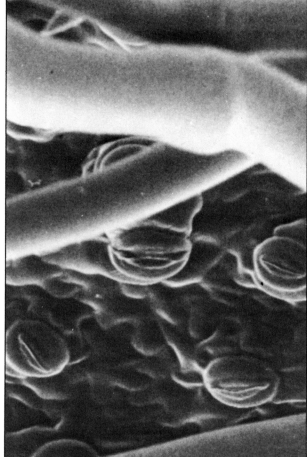

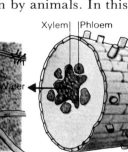

Xylem Phloem

Root hair

Water

Left: The flow of water through a plant, the transpiration stream, begins in the root. Water is taken into the root via the root hairs. From there it passes through the root tissues into the central core of xylem vessels. The xylem vessels of the root connect with those of the stem.

place. Each mitochondrion is surrounded by a double membrane. The inner membrane is folded inwards to form cristae. Respiration occurs on these cristae.
Mitosis is the process of cell division that forms new cells in an organism.

N **Nuclear membrane.** This membrane surrounds the nucleus of a cell. It is perforated by a number of pores which allow RNA to pass through. The ENDOPLASMIC RETICULUM is connected to the pores in the nuclear membrane.

Nucleolus. This is a small dense body in the nucleus of a cell. Its function is to make RIBOSOMES. A nucleolus consists of larger amounts of RNA and protein, together with a small amount of chromosomal DNA, known as the nucleolar organizer.
Nucleus. This contains the CHROMOSOMES of a cell. It is the central organizer of the cell. When a cell divides, the nucleus divides into 2 new identical nuclei.

O **Organelles** are structures in the cytoplasm of a cell that have particular

Osmosis

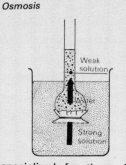

Weak solution

Water

Strong solution

specialized functions. The main organelles of a plant cell include the CHLOROPLASTS, MITOCHONDRIA, and the ENDOPLASMIC RETICULUM.

Osmosis is the passage of water through a SEMI-PERMEABLE MEMBRANE from a weaker solution to a stronger one. The pressure needed to prevent this passage of water is called the osmotic pressure, which is greater when the solution is stronger. A weak solution is said to have a high diffusion pressure. Water thus passes from a solution that has a high diffusion pressure to a solution that has a low diffusion pressure. In a laboratory experiment using 2 sugar solutions separated by a semi-permeable mem-

brane, osmosis continues until the solutions are the same strength. The root cells of plants have a lower diffusion pressure than the soil

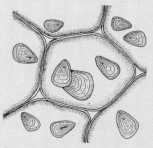

Parenchyma cells

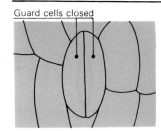

Guard cells closed

Left: The stomata of a leaf are the openings from which water escapes from the plant during transpiration. But they are also the 'breathing pores' of the plant. Generally stomata are open during the day (*below*) and are closed at night (*left*). In extreme conditions of drought the stomata may close during the day when the plant wilts.

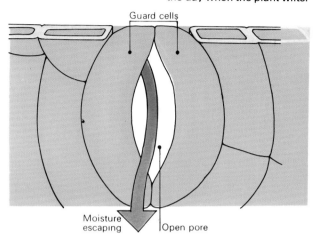

Guard cells

Moisture escaping   Open pore

Below: Photosynthesis can only take place in the light. Thus the leaves of a plant are always arranged so that they receive the maximum possible amount of light. Trees, such as these beech trees, have most of their leaves spread out sideways. The leaves are also tilted, presenting the greatest surface area to the sunlight.

energy contained in the food is passed on. Animals use the energy in movement and other activities of the body. Eventually, both animals and plants die, and bacteria use the energy as they play their part in the decay of living matter. In whatever way the energy is used it is ultimately converted to heat. This radiates out into the atmosphere, and from there is lost into space.

Thus, the secret of life on Earth is the chemical that enables photosynthesis to take place. This chemical is chlorophyll. It is a green pigment made up of complicated molecules. One essential ingredient of the chlorophyll molecule is magnesium, which must therefore be one of the minerals taken up from the soil in the transpiration stream. Chlorophyll is contained in the CHLOROPLASTS, and it is here that photosynthesis takes place.

Photosynthesis, like many of the chemical processes of life, is a long chain of chemical reactions. However, it is basically the conversion of carbon dioxide and water into sugar. The carbon dioxide is taken from the air, and oxygen is released. Some of this sugar is converted to starch by the action of ENZYMES.

---

water around them. Thus, water passes into the cells from the soil. Inside the plant, osmosis causes water to pass from one cell to another. But a cell cannot take in water indefinitely. The cell wall exerts TURGOR PRESSURE. This limits the amount of water taken in, and the internal solution does not necessarily become equal in strength to the external solution. *See also* SUCTION PRESSURE.

P **Palisade cells** form the tissue in the upper part of the leaf of a flowering

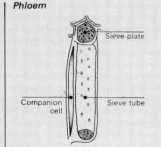

Phloem

Sieve-plate

Companion cell

Sieve tube

plant. They are elongated cells that contain large numbers of CHLOROPLASTS. Most of the photosynthesis that occurs in the plant takes

place here.

Parenchyma is a tissue that consists of thin-walled, many-sided cells. It is found in the central pith of stems, and in the cortex – the tissue that lies outside the XYLEM and PHLOEM of non-woody plants.

Phloem is the tissue that conducts food and other materials up and down the plant. Phloem tissue contains the conducting cells, or sieve tubes, together with companion cells, sclerenchyma fibres and parenchyma cells.

Plasma membrane. This

cell membrane surrounds the cytoplasm. It is SEMI-PERMEABLE and lies just inside the cell wall.

Plasmodesmata are the

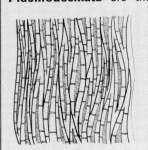

Sclerenchyma fibres

protoplasmic connections between cells. Thin strands of cytoplasm pass through pores in the cell walls.

Proteins are complicated chemical compounds composed of long chains of AMINO ACIDS. *See* RNA.

Protoplasm consists of all the contents of a cell – including the CYTOPLASM and the NUCLEUS.

R **RNA** (Ribonucleic acid) is the chemical compound in the cell concerned with making proteins. It has the same basic structure as DNA, but it contains the base

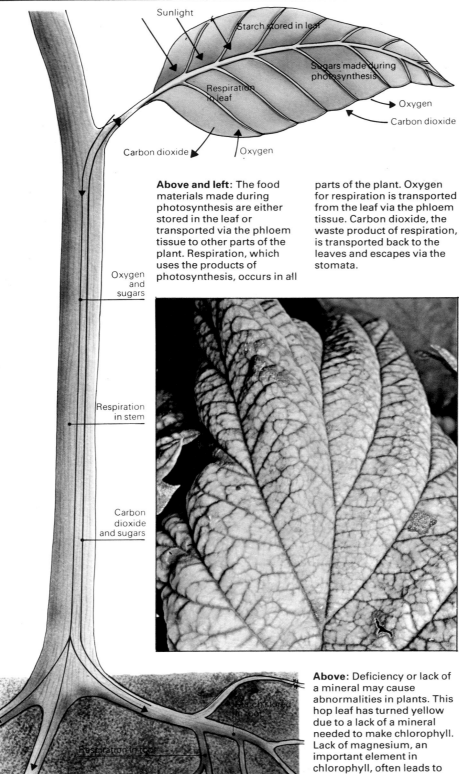

Sunlight

Starch stored in leaf

Sugars made during photosynthesis

Respiration in leaf

Oxygen

Carbon dioxide

Carbon dioxide

Oxygen

Oxygen and sugars

Respiration in stem

Carbon dioxide and sugars

Starch stored in root

Respiration in root

Above and left: The food materials made during photosynthesis are either stored in the leaf or transported via the phloem tissue to other parts of the plant. Respiration, which uses the products of photosynthesis, occurs in all parts of the plant. Oxygen for respiration is transported from the leaf via the phloem tissue. Carbon dioxide, the waste product of respiration, is transported back to the leaves and escapes via the stomata.

Above: Deficiency or lack of a mineral may cause abnormalities in plants. This hop leaf has turned yellow due to a lack of a mineral needed to make chlorophyll. Lack of magnesium, an important element in chlorophyll, often leads to such yellowing.

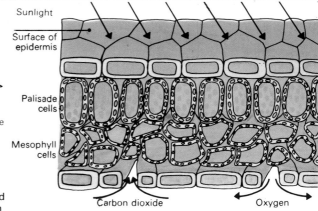

Sunlight

Surface of epidermis

Palisade cells

Mesophyll cells

Carbon dioxide

Oxygen

Above: A cross-section through a leaf. Carbon dioxide enters the leaf via the stomata and passes into the mesophyll cells via the air spaces that surround them. The dissolved carbon dioxide then diffuses into the palisade cells, which contain many chloroplasts. Here, photosynthesis takes place – assisted by sunlight.

Respiration can be regarded as the reverse of photosynthesis. Using oxygen from the air, sugar is chemically broken down by enzymes, and carbon dioxide, water and energy are released in the process. If there is not enough sugar present, some of the stored starch is converted back into sugar.

The energy released is stored in a compound called adenosine triphosphate (ATP). This is made by using the energy to add one phosphate group to adenosine diphosphate (ADP). ATP is a high energy compound. Its stored energy can be used to drive other chemical reactions. As it loses its energy, a phosphate group breaks off, and ADP is formed again.

During photosynthesis and respiration, oxygen, carbon and nitrogen are continuously being taken in and released by living organisms. There is thus a continuous cycle of events in which these elements are used.

Carbon is not used by itself, but is combined with oxygen to form carbon dioxide in the air. This is used by plants during photosynthesis to make CARBOHYDRATE chemicals, such as sugars, starch and cellulose. These are passed on to animals when the plants are eaten. Carbon dioxide is released back into the air in two ways. During respiration sugars are broken down to provide energy and carbon dioxide is given off in the process. Also, when plants and animals die, bacteria break down the carbohydrates into carbon dioxide and water.

uracil instead of thymine. Also, RNA molecules only consist of single strands. The 2 main forms of RNA concerned in protein synthesis are messenger RNA and transfer RNA. When a particular protein is to be made, a short strand of messenger RNA is formed by the DNA of the nucleus. This RNA contains certain 'orders' in the form of coded bases. The messenger RNA then moves into the cytoplasm, where it forms transfer RNA. This then picks up free AMINO ACIDS according to the 'orders' passed on by the messenger RNA. Meanwhile, the messenger RNA has positioned itself on the RIBOSOMES of the endoplasmic reticulum. The transfer RNA carries the amino acids to the messenger RNA, and the amino acids join up. The sequence in which the amino acids are joined together is controlled by the sequence of bases on the messenger RNA. Thus, the particular protein 'ordered' by the nucleus is made.

Ribosomes are granular bodies in the cytoplasm of a cell. Many ribosomes are attached to the surface of the ENDOPLASMIC RETICULUM. They are composed of RNA and protein and are made by the NUCLEOLUS. Sometimes they are linked together in long chains called polyribosomes. They are probably linked by a molecule of messenger RNA.

Root hairs are extensions of the epidermal cells of a root. The extension takes the form of a tube which grows out of the cell. Root hairs have very thin walls, and thus allow water to pass easily from the soil into the cells.

Root pressure is the pressure that causes water to pass from root cells into the XYLEM cells of the root. It is due to the fact that OSMOSIS occurs in the root cells. Differences in osmotic pressure

Stone cells

Sunlight in Wyre Forest

Oxygen in the air is used up during respiration. But plants also give off oxygen during photosynthesis. In fact, plants are essential for maintaining the oxygen content of the air.

Nitrogen is an extremely important element for living organisms. It is an essential part of AMINO ACIDS, which are the building blocks of proteins. There are large quantities (78 per cent) of nitrogen in the air, but unfortunately, this cannot be used by most plants. However, some organisms can 'fix' nitrogen and convert it directly into amino acids. In some cases these organisms have a symbiotic relationship with plant roots which is mutually beneficial. For example, the root nodules of pea plants contain a bacterium called *Rhizobium* which can 'fix' nitrogen inside the root nodules.

All the other elements that plants need are obtained as minerals from the soil. In addition to nitrogen, the most important elements are potassium, calcium, phosphorus, magnesium, sulphur and iron. Although they are essential, they are only needed in small quantities. For example, the solid parts of a plant (i.e. excluding the water) only contain about three and a half per cent potassium. But without potassium a plant cannot survive.

Other elements, called trace elements, are needed in even smaller quantities — less than 0.0001 per cent. These are copper, manganese, zinc, molybdenum and boron. Even so they are still essential. For example, lack of zinc causes the leaves of a plant to become deformed.

Plant pigments

The countryside is largely green, and this is due

Below: Oxygen, carbon and nitrogen are continually being recycled. Carbon is present as carbon dioxide in the air and as the many carbon compounds that make up the bodies of animals and plants. Carbon dioxide is taken up during photosynthesis. It is released during respiration, when plants and animals die and decompose, and when fuels are burned. Oxygen in the air is used up during respiration and burning, and is given off during photosynthesis. Nitrogen is mostly taken up by plants in the form of nitrates, although some bacteria can 'fix' nitrogen directly from the air.

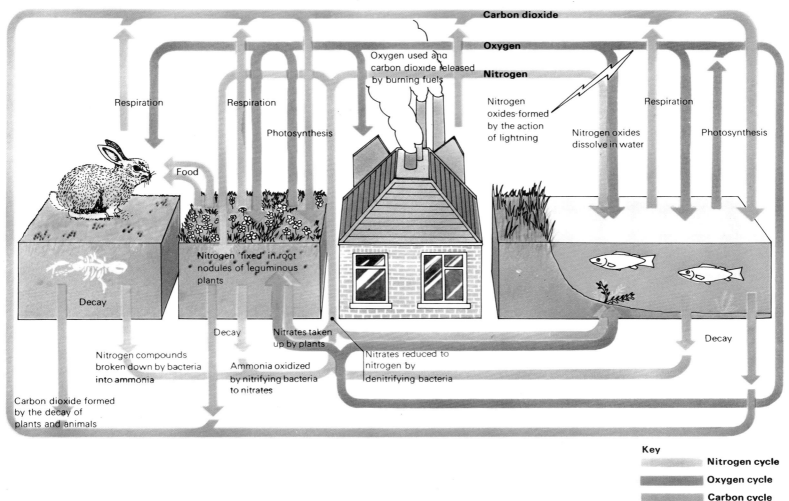

Carbon dioxide

Oxygen

Nitrogen

Oxygen used and carbon dioxide released by burning fuels

Respiration

Respiration

Photosynthesis

Nitrogen oxides formed by the action of lightning

Nitrogen oxides dissolve in water

Respiration

Photosynthesis

Food

Nitrogen 'fixed' in root nodules of leguminous plants

Decay

Decay

Nitrates taken up by plants

Decay

Nitrogen compounds broken down by bacteria into ammonia

Ammonia oxidized by nitrifying bacteria to nitrates

Nitrates reduced to nitrogen by denitrifying bacteria

Carbon dioxide formed by the decay of plants and animals

Key

Nitrogen cycle

Oxygen cycle

Carbon cycle

---

are maintained by ions being actively transported by a special mechanism across cell membranes. When transpiration is slow, root pressure may cause the phenomenon of guttation. Water exudes through special glands on the edges of leaves.

S **Sclerenchyma** is a tissue that consists of stiff, non-living cells, heavily thickened with cellulose, lignin, or both. Sclerenchyma fibres are long, thin cells, pointed at both ends. They are found in many parts of plants. The cellulose fibres of the flax plant (*Linum usitatissimum*) are used for making linen, and the 'lignified fibres of hemp (*Cannabis sativa*) are used for making rope.

Semi-permeable membrane is a membrane that allows some substances to diffuse through it but not others. The plasma membrane of a cell is a semipermeable membrane that allows water to pass through, but not substances such as mineral ions and proteins dissolved in water.

Starch is a CARBOHYDRATE. It is also a polysaccharide, i.e. its molecules are made up of many sugar units. Starch is the main food storage compound used by plants. It is found in large quantities in seeds and storage organs such as potato tubers.

Stomata (*singular:* stoma) are the 'breathing pores' on the undersides of the leaves of plants. A stoma consists of 2 guard cells that are shaped like kidneys. By altering the TURGOR PRESSURE inside the guard cells, the plant can open or close them. Carbon dioxide and oxygen pass in and out of the stomata during photosynthesis and respiration. During transpiration, water is lost by evaporation from the stomata.

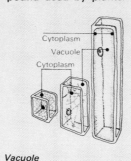

Cytoplasm

Vacuole

Cytoplasm

Vacuole

Antirrhinum

to the chlorophyll pigment present in most plants. But there are also other plant pigments, and some of these give us the other brilliant colours of nature.

The xanthophylls and carotenes are a group of pigments ranging in colour from yellows to reds. Like chlorophyll they are photosynthetic pigments and are found in the leaves. In the autumn the chlorophyll of a deciduous tree is withdrawn from the leaves. The xanthophylls and carotenes that remain give the leaves their familiar yellow-brown autumn colour. The same pigments give the colour to apples, carrots, tomatoes and brown algae.

The anthocyanin pigments provide the most spectacular of the plant colours — the bright violets, blues and reds. They are mostly found in flowers, but some fruits also contain anthocyanins.

The last group of pigments are phycoerythrin and phycocyanin. They only occur in certain groups of algae, such as the blue-green algae and the red algae.

Below: The reds, browns and yellows of autumn leaves are due to the presence of xanthophylls and carotenes.

Above: The brilliant blue colour of gentian violets is due to the presence of one of the anthocyanin pigments.

Right: The familiar bright yellow colour of daffodils is produced by xanthophyll and carotene pigments.

Stone cells, or sclereids, are irregularly shaped cells. They are common in seed coats and fruits, such as pears and nuts.

Suberin is a waterproof substance made up of a number of carbon compounds. It is found in the walls of cork cells. The chemicals involved in suberin are all formed from compounds called fatty acids.

Suction pressure is the capacity of a cell to take in water. It is equal to the OSMOTIC PRESSURE less the TURGOR PRESSURE. Suction

Xylem

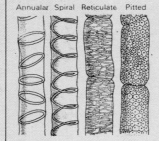

Annualar Spiral Reticulate Pitted

pressure is not an active force of the cell. Water is pushed rather than sucked in. Inside a plant, water moves from one cell to

another until the concentration of water in the cells is equal.

Sugars are carbohydrates that have a sweet taste. They are divided into 2 groups. Monosaccharides are the simple sugars whose molecules have 5 or 6 carbon atoms. Examples include ribose, an important constituent of DNA and RNA, and glucose, which is made by plants during photosynthesis. Disaccharides are sugars whose molecules are made up of 2 monosaccharide molecules. An example is sucrose.

Transpiration is the loss of water by evaporation from the stomata of a plant. Some water is also lost through the cuticle of the epidermis.

Transpiration stream covers the flow of water from the ROOT HAIRS, through the XYLEM vessels, to the STOMATA of a plant.

Turgor pressure (wall pressure) is the pressure exerted by the cell wall that tends to prevent water entering the cell by OSMOSIS. Turgor is a term used to describe the state of the cell when the cell wall is stretch-

ed as far as possible.

Vacuole. This is the fluid-filled space enclosed by the cytoplasm of the cell. Its presence is due to the fact that when cells grow, the amount of cytoplasm does not increase, so it becomes stretched round the inside of the cell wall.

Xylem is the tissue that conducts water and minerals up the plant. It consists of xylem vessels, or tracheids, SCLERENCHYMA fibres, and PARENCHYMA cells.

We often associate bacteria with disease, but they also perform the vital role of breaking down all dead organisms. Bacteria and many algae are microscopic, single-celled plants, but algae also include giant seaweeds.

Bacteria and Algae

Right: Some forms of bacteria.
1. Bacilli have rod-shaped cells. **2.** Cocci have spherical cells. Some cocci occur in clusters, such as *Staphylococcus*. **3.** Spirillar bacteria have curved or twisted rod-shaped cells, often with flagella. **4.** *Streptococcus* has long chains of spherical cells.

Below: Bacteria are used in the making of cheeses. One type of bacterium begins the process by causing the milk to curdle. Other types of bacteria make acid or digest fats or proteins. Different types of cheese are made by varying the extent to which each type of bacterium plays a part.

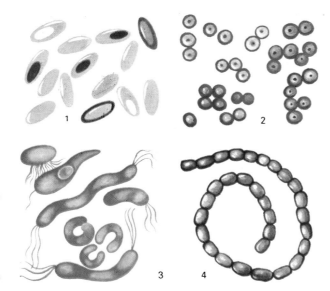

Bacteria are minute single-celled organisms. Many scientists include them in the plant kingdom, but they are not like other plants. Their cells have no nuclei, and their cell walls are not made of cellulose. Instead, they are made up of a number of substances, including proteins and fats. Some bacteria can move, others cannot. Some need oxygen to survive, but others are poisoned by oxygen.

A few bacteria make their own food by photosynthesis, but most of them live by breaking down dead plant and animal material. Such bacteria are essential for life to continue. When they break down materials, they make simple chemicals that can be used again by other plants. Thus, there is no waste in the living world.

Some of these bacteria are particularly useful to us. In the treatment of sewage, bacteria are allowed to act on the raw sewage and make it harmless. We also use bacteria to make silage and garden compost by rotting down plant material.

Some plants and animals use bacteria more directly. Many plant-eating animals, such as rabbits, horses and cows, have bacteria in their digestive systems. The presence of these bacteria is essential for the breakdown and digestion of cellulose, which the animals cannot do by themselves. Some flowering plants, such as clover and other members of the pea family, have bacteria in their roots. These bacteria help the plants by converting, or fixing, nitrogen into a form that plants can use.

Many bacteria have less desirable effects. For example, they cause food to decay. Also, the presence of some bacteria in food causes food poisoning. Many other diseases, such as typhoid fever, tuberculosis and cholera are caused by bacteria. Plants, too, may be infected, as for instance, soft rot in carrots and fire blight in

---

Reference

A **Anisogamy** is a type of sexual reproduction in which 2 unlike sex cells (gametes) fuse. The larger of the 2 cells (the 'female') is able to move by means of one or more FLAGELLA. *See also* ISOGAMY, OOGAMY.
Antheridium. This is the male sex organ of a lower plant, e.g. some brown algae, the liverworts, mosses and ferns. Inside an antheridium the male sex cells, or antherozoids, are

formed.
Asexual reproduction is the way in which an organism reproduces itself without involving sex cells.

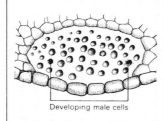
Developing male cells
Antheridium

B **Bacillariophytes,** see DIATOMS.
Bacteria are a group of single-celled organisms that have no definite nucleus (their DNA is distributed throughout their cells), and no chlorophyll. They are usually classified as plants, but sometimes they are put into a third kingdom, Protista, which includes the protozoa (single-celled animals), single-celled algae and fungi.

C **Chlamydomonas** is a genus of single-celled algae belonging to the

Diatoms

Chlorophytes. Members of this genus have 2 flagella. Most of the cell is occupied by a single, basin-shaped chloroplast.
Chrysophytes are a group

of single-celled, or colonial, algae. They are golden brown.
Cyanophytes are the blue-green algae. They have no definite nuclei (their DNA is distributed throughout their cells). They exist in both single-celled and filamentous forms, and they are mostly found in fresh water. They contain chlorophyll and phycoerythrin and phycobilin. These pigments give them their characteristic blue-green colour.

D **Decay,** see PUTREFACTION.

pears. These diseases are slowly disappearing.

The closest relatives of the bacteria are the blue-green algae. Like bacteria they do not have cellulose cell walls, and their cells do not have definite nuclei. In addition, many blue-green algae are able to fix nitrogen. However, blue-green algae have one main thing in common with other plants. they contain chlorophyll and can therefore make their own food.

Most blue-green algae are composed of strings, or filaments, of cells. These are surrounded by a slimy material called mucilage. Blue-green algae live in ponds and streams, in the sea, and even in hot springs. Sometimes they cause problems by growing in drinking water supplies.

Algae

Algae vary from single-celled plants to large many-celled seaweeds. At first glance, such

Right: Some forms of algae. *Euglena* is a green alga that swims by using its whip-like hair, or flagellum. *Nostoc* is a blue-green alga. Its long chains of cells are contained within a slimy sheath. *Coscinodiscus* is a diatom. Its hard cell wall is made of silica and is made in 2 parts – like a pill box. *Cladophora* is a branched filamentous alga. *Fucus* is a brown alga, familiar as the seaweed found on many shores.

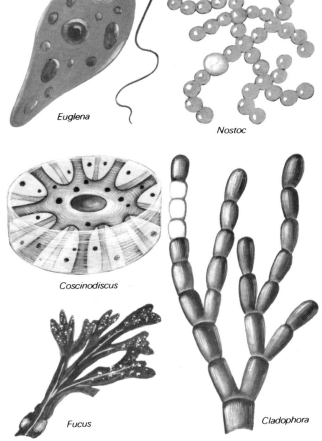

Euglena

Nostoc

Coscinodiscus

Fucus

Cladophora

Left: During the process of sewage treatment bacteria are used to destroy the waste matter. Air is bubbled through the sewage to provide the oxygen necessary for this.

plants would seem to have little in common. However, they are all simple plants with no roots or stems, and they all live permanently in water, between the high and low tide marks, or on damp walls and trees.

Many single-celled algae float freely in the water, together with many tiny animals. This drifting mass of minute animals and plants is called plankton. The animals (zooplankton) feed on the plants (phytoplankton), and both are eaten by larger plankton-eating animals, such as fish and whales.

Among the single-celled algae are simple round forms, such as *Pleurococcus*, which grows on damp tree trunks. Other single-celled algae have complicated shapes, such as *Ceratium*, an algae that lives in the sea. The DIATOMS are a group of algae that have silica in their cell walls. Many

---

Diatoms are a group of single-celled algae, also known as Bacillariophytes. They have cell walls that contain silica, and they are golden brown in colour.

E **Euglena** is a genus of single-celled algae belonging to the Euglenophyta. Members of this genus have a long flagellum that emerges from a gullet at the front end of the cell. A disk – or star-shaped chloroplast is also present, and thus *Euglena* species make their own food by photosynthesis. Individuals that grow in the

dark lose their chloroplasts and feed by taking in organic material – i.e. in the same way as animals.

Euglenophytes are a group of single-celled algae. They have 1, 2 or 3 flagella. They are green in colour and are generally regarded as plants. However, they do have some animal characteristics (*see* EUGLENA).

F **Fertilization** is the fusion of 2 sex cells. The cell that results from this fusion is called a *zygote*.
Flagellum (*plural:* flagella) is a whip-like organ pos-

sessed by algae and protozoans (single–celled animals). By lashing its flagellum an organism is able to swim through the water.

Bacteria culture

Food poisoning occurs when certain chemicals or bacteria are eaten with food. Bacteria that cause food poisoning include salmonellae, staphylococci and clostridia.
Fucus is a genus of brown seaweeds belonging to the Phaeophytes. There are several well–known species, including *F. vesiculosus* (bladder wrack), *F. serratus* (serrated wrack) and *F. spiralis* (twisted wrack).

G **Gametes** are sex cells. During the process of sexual reproduction 2 ga-

Laminaria

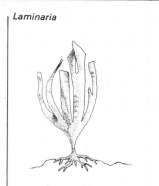

metes fuse to form a zygote.

I **Isogamy** is a type of sexual reproduction in

diatoms, too, have beautifully sculptured shapes. A large number of algae, such as CHLAMYDOMONAS and EUGLENA, have whip-like organs called flagella. By using these they are able to move from place to place — usually a characteristic of animals. In fact, scientists have argued about whether *Euglena* is a plant or an animal, as it not only has a flagellum, but also has other animal-like characteristics.

The simplest type of many-celled alga is called a colony. This is just a group of cells that all work together. For example, *Volvox* consists of a round, hollow ball of cells; each cell is like a single *Chlamydomonas* individual.

Many algae consist of long filaments (threads) of cells joined end to end. The simplest of these include ULOTHRIX and SPIROGYRA, which both have long, unbranched filaments. More complicated types have branched filaments. *Cladophora* has a tree-like system of branches, and it is anchored to the ground by a tiny root-like structure. *Stigeoclonium* has two sets of branches. One reaches upwards like a tree; the other spreads over the ground.

Most of these algae can only be seen clearly with a microscope, but there are a number of larger types. The STONEWORTS, such as *Chara*, are a freshwater group that resemble higher plants. They have 'stems' with rings of 'branches' around them. Most of the larger algae grow in the sea, and we know them as seaweeds. SEA LETTUCE is a green seaweed that consists of a flat

Below: Some tropical species of oarweeds may reach 200 metres in length. They are anchored to the rocks by holdfasts or root substitutes.

Above: The green colour of the bark of this beech tree is due to the presence of the round, single-celled, green alga *Pleurococcus*.

Below: *Volvox* is a colonial green alga in which many cells are arranged to form a hollow ball. These 4 colonies each have daughter colonies inside.

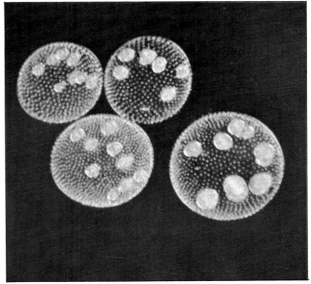

which 2 identical sex cells (gametes) fuse. *See also* ANISOGAMY, OOGAMY.

L Laminaria is a genus of brown seaweeds belonging to the PHAEOPHYTES. They have large flattened fronds. Because of their appearance, they are often called oar weeds.

N Nitrogen-fixing bacteria are those that can convert nitrogen into nitrogen-containing salts that can be used by plants. They include *Rhizobium* which is found in the root nodules of leguminous plants, and *Azotobacter*, which lives free in the soil.

O Oogamy is a type of sexual reproduction in which a swimming male sex cell (antherozoid) fuses with a large, stationary female sex cell (oosphere).
Oogonium is the female sex organ of some brown algae and fungi. Inside an oogonium, one or more female sex cells, or oospheres, are formed.

P Phaeophytes are a group of brown sea algae. Most of them, including FUCUS and LAMINARIA, grow in-between the high and low tide marks.
Pleurococcus is a genus of

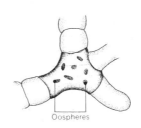

Oogonium

algae belonging to the CHLOROPHYTES. It is one of the simplest algae, consisting of a single round cell without any flagella. Its only method of reproduction is by simple division.
Putrefaction is the decomposition by BACTERIA of plant or animal material that contains protein. During this process the proteins are broken down into various foul-smelling chemicals.
Pyrrophytes are a group of single-celled algae that range in colour from yellow through green to dark brown. This group includes

Sea lettuce

the dinoflagellates, which often glow in the sea.

R Rhodophytes are a group of red algae. They

sheet of cells. Red seaweeds (RHODOPHYTES) often grow to over 30 centimetres in length. Many of them consist of a complicated and delicate cell arrangement. However, the brown seaweeds are probably the most familiar. This group includes the largest of all the algae; some of them can be over 100 metres long. They include the wracks and kelps. Many of these anchor themselves to the rock by structures called holdfasts. Some seaweeds float freely.

How algae reproduce

The simplest forms of reproduction do not involve any sex cells. If parts of a filamentous algae break off, they continue to grow and thus form new individuals. Many single-celled algae reproduce by simply dividing into two. A more complicated type of asexual ('without sex') reproduction involves the formation of swimming cells called zoospores. When these are released,

they can swim to a new site and then grow.

Sexual reproduction involves sex cells. Some algae produce only one kind of swimming sex cell. In such cases the two identical sex cells fuse together, and the resulting cell, or zygote, grows into a new plant. A slightly more advanced form of sexual reproduction occurs when unlike sex cells are formed. The larger of the two (the 'female') swims to meet the smaller sex cell (the 'male') with which it then fuses.

The most advanced type of reproduction found in the algae involves a swimming male sex cell, called an antherozoid, and a large, non-swimming female sex cell, called an oosphere. Antherozoids are produced in a special sex organ called an antheridium. An oosphere is produced in an oogonium. The antherozoids swim to the oosphere and fusion, or fertilization, takes place inside the oogonium. The zygote is released and eventually grows into a new plant.

Below: In asexual reproduction of *Chlamydomonas* (top), an individual simply divides into 2 inside its cell wall. Sexual reproduction in *Chlamydomonas* (bottom) involves the formation of sex cells.

Right: Asexual reproduction in *Spirogyra* simply involves breaking the filament. New cells are then added on to the 2 halves by mitosis.

Below: A single cell of the filamentous alga *Spirogyra*. Its spiral chloroplast is unique to the plant kingdom.

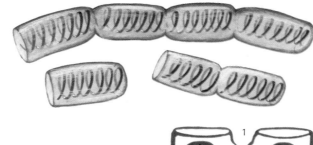

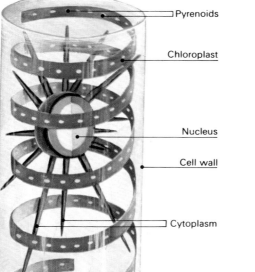

- Pyrenoids
- Chloroplast
- Nucleus
- Cell wall
- Cytoplasm
- Vacuole

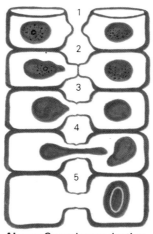

Above: Sexual reproduction, or conjugation in *Spirogyra*. **1.** Cells of neighbouring filaments produce outgrowths. **2.** The outgrowths join to form a tunnel. **3.** One nucleus begins to migrate through the tunnel. **4.** The 2 nuclei fuse. **5.** A zygote is formed.

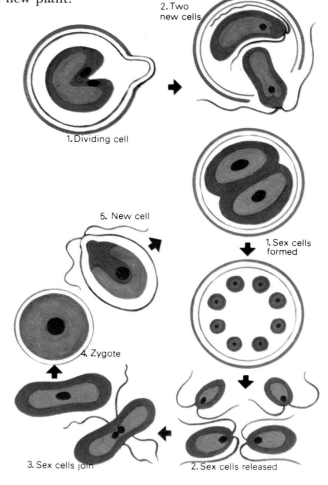

2. Two new cells

1. Dividing cell

5. New cell

1. Sex cells formed

4. Zygote

3. Sex cells join

2. Sex cells released

---

are many-celled and often have delicate shapes. They mostly grow in the sea, and their red colour is due to pigments that allow photosynthesis to occur in deep water where the light is dim.

S **Sea lettuce** (*Ulva lactuca*) is a species of algae belonging to the CHLOROPHYTES. It grows in the sea, between high and low tide marks, and it consists of a flat sheet of cells attached to a rock by a short stalk.
Sexual reproduction is the way in which an organism reproduces itself when sex

cells (gametes) are involved.
Spirogyra is a genus of

Clover plant with root nodules

algae belonging to the CHLOROPHYTES. They are unbranched, filamentous algae that grow in fresh water. Each cell has an unusual chloroplast which forms a spiral that runs the length of the cell.
Stigeoclonium is a genus of algae belonging to the CHLOROPHYTES. They are branched, filamentous algae, and each branch ends in a tapering cell.
Stoneworts are a group of green algae that are totally unlike any other group. They have distinct 'stems', and at intervals along the stems

they have rings of small 'branches'. There are only 8 types of stonewort, all of which are found in fresh water; the 2 most common are *Chara* and *Nitella*.

U **Ulothrix** is a genus of algae belonging to the Chlorophytes. They are unbranched, filamentous algae, with ring-shaped chloroplasts in their cells.
Ulva lactuca, see SEA LETTUCE.

V **Volvox** is a genus of colonial algae belonging to the Chlorophytes group.

They consist of hollow balls of CHLAMYDOMONAS-like cells.

W **Wracks,** see FUCUS.

X **Xanthophytes** are a group of algae that are yellow-green in colour. They are mostly single-celled plants that grow in fresh water or in damp soil.

Z **Zygote,** see FERTILIZATION.

The familiar mushroom and the tiny pin mould on stale bread belong to the same large division within the plant kingdom – the fungi. Sometimes a fungus lives in a strange partnership with algae, which is called a lichen.

Fungi and Lichens

Above: *Mucor,* the pin mould, is often found growing on stale bread. The name pin mould comes from the pin-like structures called sporangia that contain the asexual spores.
Left: The sexual reproduction of *Mucor* involves nuclei from 2 separate hyphae (threads). They fuse and a tough zygospore is formed. When conditions are right this germinates to form a new mass of hyphae, or mycelium.

Fungi are plants that do not contain chlorophyll (green pigment). As a result they cannot make their own food; they have to use sources of ready-made food. There are over 50,000 kinds of fungi, and they include the moulds, rusts, yeasts and toadstools, as well as several other groups.

Some fungi live on decaying plant or animal tissue. They are called saprophytes and they produce enzymes that break down the chemicals of the tissue. The fungi can then absorb the broken down chemicals. Other fungi are parasites – that is they live on other organisms without benefiting their hosts in any way. Some parasites do not harm their hosts, but others destroy their hosts completely.

A few fungi are single-celled, but most of them have a plant body called a mycelium, which is made up of long threads, or hyphae. There are

Above: Potato blight is a disease caused by the fungus *Phytophathora infestans.* The fungus attacks the stem and leaves causing them to die.

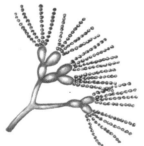

Above: *Penicillium* is a fungus that produces long chains of asexual spores, or conidia.

three main groups of fungi, and the main difference between them is in the way that the members of each group reproduce.

Phycomycetes – simple fungi
The simplest fungi include the PIN MOULDS, downy mildews and potato blight. Species of pin mould can often be found growing on stale bread and other foods. The mycelium spreads over the surface of the food and produces upright branches with tiny pin heads. These heads are spore-containers called sporangia. Pin moulds also have a form of sexual reproduction. Two chemically different hyphae produce outgrowths that join up to produce a hard black object called a zygospore. From this grows a single thread, which produces a sporangium at the tip.

Downy mildews are parasites, and various

Right: A rust fungus on the leaves of coltsfoot. The rust-red patches are where the spores of the fungus are produced.
Left: The fruiting body of a Morel (*Morchella esculenta*). This is an edible fungus that grows on rich soil in spring. It is an ascomycete and its spores are produced in the hollows on the 'cap'.

species attack a number of flowering plants, including onions, cabbages, tobacco and maize. The fungus grows in the spaces between the cells of the host, and inside the cells it produces club-shaped organs that absorb food from the host.

POTATO BLIGHT is a serious disease of potatoes. During the winter it lives inside potato tubers. In the spring the mycelium grows into the potato shoots and can completely ruin a potato crop. The fungus is spread by spores, which are produced on the surface of the leaves of an infected host plant.

Ascomycetes – more advanced fungi

Ascomycetes get their name from the fact that during sexual reproduction they produce spores in structures called asci (*see* ASCUS). Many ascomycetes produce their asci in fruiting bodies that can be seen without a microscope. *Peziza* has a cup-shaped fruiting body. Morels have large, stalked fruiting bodies.

Many ascomycetes are parasites, and their fruiting bodies are small. The powdery mildews attack a wide range of plants, including gooseberry bushes, strawberry plants, oak trees and chrysanthemums. They produce round fruiting bodies in which the asci are completely enclosed.

The ERGOT OF RYE is a parasite that has flask-shaped fruiting bodies. These grow from small, black masses of hyphae, called ergots, which replace some of the grains of rye. Ergots can be used to produce the drug ergotin, which is used to stop bleeding. If they are ground up into flour with the rest of the grain, they can cause a serious disease in man and animals called ergotism.

Some ascomycetes are single-celled fungi. The alcohol-producing yeasts are examples of this type. Like other ascomycetes they can reproduce by forming asci, but they mainly reproduce by

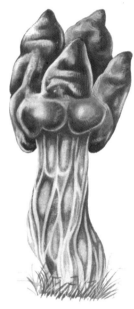

Above: *Helvella lacunosa* is a woodland fungus that produces its fruiting bodies in autumn. It is an ascomycete and its spores are formed on the grey 'cap'.

budding. Each cell divides into two, and in this way the fungus grows long chains of cells. These fungi can live on sugar and convert it into ethyl alcohol and carbon dioxide, and they are therefore used in the fermentation of wines and beers. Other single-celled ascomycetes include peach leaf curl and Dutch elm disease.

Many species of PENICILLIUM, the fungus from which we get penicillin, do not produce asci – or at least they have never been discovered. However, we know that they are ascomycetes because of the detailed structure of their mycelia. *Penicillium* produces asexual spores called conidia on the head of a club-shaped structure called a conidiophore. There are several ascomycetes whose asci have not yet been discovered, and they are often classed in a separate group called the *Fungi Imperfecti*.

Basidiomycetes – the most advanced fungi

Basidiomycetes produce spores on structures called basidia (*see* BASIDIUM). Many basidiomycetes have large fruiting bodies, and this group includes the MUSHROOMS and TOADSTOOLS.

The rusts and smuts are parasitic basidiomy-

---

species is *B. edulis* which has a brown cap and greenish-yellow pores. The only poisonous species is the devil's boletus (*B. satanus*), which can easily be recognized by its red pores and stipe.

C **Chanterelle** (*Canth-erellus cibarius*) is a funnel-shaped toadstool that is egg-yolk yellow in colour and has a faint smell of apricots. It grows in all woods in summer and autumn. It is difficult to find, but this excellently flavoured fungus is worth looking for.

Coprinus is a large genus of toadstools that includes the ink caps. The easiest to identify is the shaggy ink cap (*C. comatus*), which has a tall, white, shaggy cap and pink gills that turn black and disintegrate as the spores ripen. It is edible, but only the young fruiting bodies are recommended.

D **Death cap,** see AMANITA.
Dry rot (*Merulius lacrymans*) is a soft, jelly-like fungus that lives on wood. It needs damp conditions in order to start growing, but once established it can make

water from organic chemicals. It has thick, water-conducting hyphae and it can penetrate brick and stone in order to reach more

Common puff balls

wood supplies.

E **Ergot of rye** (*Claviceps purpurea*) is a parasitic ASCOMYCETE that replaces the

grains of rye with small, black ergots. It also attacks other grasses.

F **Fairy ring champignon** (*Marasmius oreades*). A small toadstool that forms clumps and rings in grassland. It is beige in colour, with a slightly pointed cap, and well separated gills. It is edible, and is recommended for drying.
Fly agaric, see AMANITA.

G **Giant puff ball** (*Calvitia gigantea*). A large, white, round fungus that grows in grassland. It may

cetes with relatively small fruiting bodies. They cause serious diseases in grain crops, such as wheat, barley, oats and maize.

The jelly fungi, such as the JEW'S EAR FUNGUS, have soft, clammy fruiting bodies, usually attached to trees. Some of them are parasites and grow on living trees, others are saprophytes and grow on dead wood.

Another group of basidiomycetes produce their spores in an enclosed fruiting body, which only opens when the spores are ripe. These include puff balls and giant puff balls. The STINKHORN is a special member of this group.

Toadstools and bracket fungi are a familiar sight in fields and woods. What you see is only the fruiting body; the main body of the fungus, the mycelium, is in the soil or wood from which the fruiting body is growing.

A toadstool consists of a stalk and an umbrella-shaped cap. The underside of the cap may be covered with gills that radiate outwards like the spokes of a wheel. These are lined with spore-bearing basidia. Some toadstools have pores lined with basidia, and a few types have spines instead of gills. Bracket fungi are similar

Left: The fly agaric *(Amanita muscaria)* is a colourful, but poisonous, toadstool found in pine and birch woods in autumn.

Above: The dryad's saddle fungus *(Polyporous squamosus)* is a bracket fungus found on elm, oak, beech and other trees in spring and summer.

Below: The orange peel fungus *(Peziza aurantia)* is an ascomycete cup fungus. Its spores are produced inside the cups.

grow larger than a man's head, and it is sometimes known as the *Tête de mort* (head of death). It is edible, and is best fried.

H **Honey fungus** *(Armillaria mellea)* is a parasitic fungus that produces toadstools at the base of its host tree. It can be a serious parasite of fruit trees, but the toadstools are good to eat.

I **Ink caps,** see COPRINUS.

J **Jew's ear fungus** *(Auricularia auricula)*. This is a

jelly fungus with brownish-purple, wrinkled fruiting bodies that sometimes resemble ears. In Britain it is found on elder trees.

Honey fungus

L **Lichens** are symbiotic associations of algae and fungi. A typical lichen consists of 3 layers. The upper layer is made up of densely interwoven fungal hyphae, forming an almost solid tissue. In the middle layer the fungal hyphae are more loosely interwoven. In the upper part of this layer are scattered groups of algal cells. The lower layer again consists of a mass of fungal hyphae, and small, root-like hyphae project from the lower surface. These hold the lichen firmly onto the rock or bark.

M **Morel** *(Morchella esculenta)* is a stalked ASCOMYCETE. The surface of the

Cladonia (lichen)

'cap' is covered in small ridges, which separate many pits. The spores are formed in these pits. Morels are excellent to eat.

Mucor, see PIN MOULD.

Mushroom is a name used to describe the fruiting bodies of a number of BASIDIOMYCETES and a few ASCOMYCETES (e.g. morels). However, it is most commonly used to describe the fruiting bodies of the edible species of AGARICUS, such as the field mushroom. There is no clear distinction between a mushroom and a TOADSTOOL.

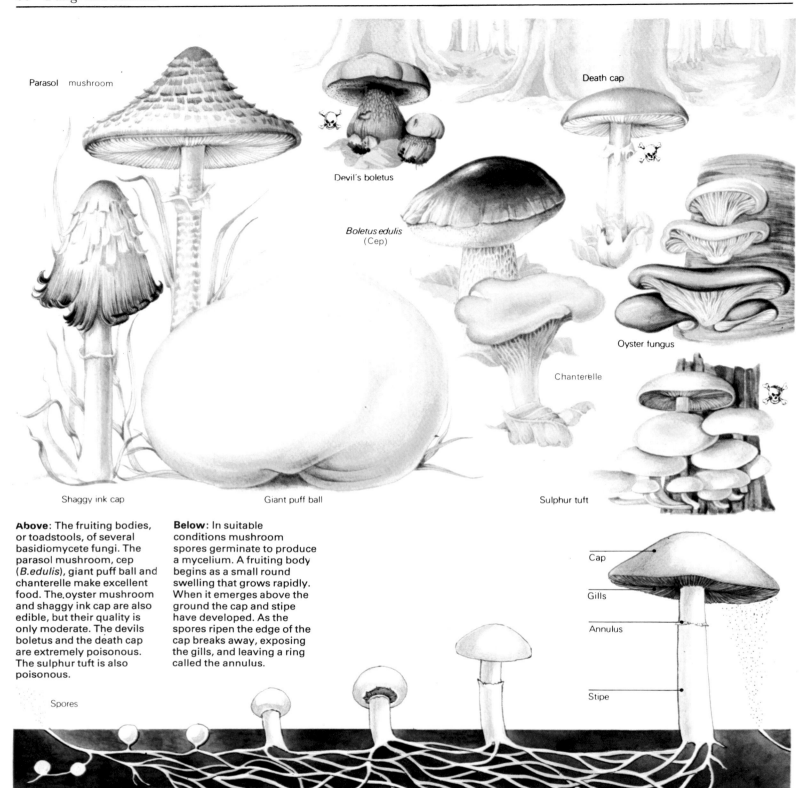

Parasol mushroom

Devil's boletus

Death cap

Boletus edulis
(Cep)

Oyster fungus

Chanterelle

Shaggy ink cap

Giant puff ball

Sulphur tuft

Cap

Gills

Annulus

Stipe

Spores

Above: The fruiting bodies, or toadstools, of several basidiomycete fungi. The parasol mushroom, cep (*B.edulis*), giant puff ball and chanterelle make excellent food. The oyster mushroom and shaggy ink cap are also edible, but their quality is only moderate. The devils boletus and the death cap are extremely poisonous. The sulphur tuft is also poisonous.

Below: In suitable conditions mushroom spores germinate to produce a mycelium. A fruiting body begins as a small round swelling that grows rapidly. When it emerges above the ground the cap and stipe have developed. As the spores ripen the edge of the cap breaks away, exposing the gills, and leaving a ring called the annulus.

O **Oyster mushroom** (*Pleurotus ostreatus*) is an edible bracket fungus found on beech and other trees. The upper surface is bluish-grey when the fruiting body is young, but turns brown as it gets older.

P **Parasite** is an organism that lives on another organism (the host) from which it obtains food. The host does not benefit, and in some cases may be destroyed. There are many parasitic fungi, such as HONEY FUNGUS and POTATO BLIGHT.
Parasol mushroom (*Lepio-*

ta procera) is an edible toadstool that grows in grassy clearings or at the edges of woods. The cap is light brown and is covered in brown scales. The stipe is striped brown and has a loose, double collar-like ring. The shaggy parasol (*L. rhacodes*) is also edible. It is similar to the parasol mushroom, but it has a smooth stipe without stripes.
Penicillium is a genus of ASCOMYCETES. Some species cause disease in weak or dormant parts of plants, such as over-ripe fruits and bulbs. Others are used to

make antibiotics, such as pencillin. Many are used in cheese, such as *P. roqueforti* and *P. camemberti*.
Phycomycetes are fungi

Young parasol mushrooms

whose hyphae do not have crosswalls. These include the PIN MOULDS.
Pin moulds (*Mucor*) are a genus of phycomycetes most of which are SAPROPHYTES. The name is derived from their pin-like structures in which asexual spores are produced.
Polyporus is a genus of bracket fungi that produce their basidia in pores. There are many species, including the dryad's saddle (*P. squamosus*), which is an edible fungus.
Potato blight (*Phytophthora infestans*) is a species

of phycomycetes that cause late blight in potatoes. It can be controlled by spraying the plants with Bordeaux mixture, which includes copper sulphate. If used in sufficient quantities this prevents germination of the spores. It does not affect the potato plant.
Puff balls (*Lycoperdon*) are basidiomycetes with small, club-shaped fruiting bodies in which the spores are completely enclosed as they ripen. The young fruiting bodies are white and they are edible. Older specimens turn brown. *Calavatia ex-*

to toadstools, except that they have no stalk; the cap of the fruiting body grows directly out of the tree in which the mycelium is growing.

Toadstools can be found growing almost everywhere. Many grow in fields, and probably the best-known species are those of the genus AGARICUS, which includes the tasty field mushroom. Many other kinds can also be eaten, *but some are poisonous, and you must take great care to identify all toadstools correctly before eating them.* If there is one you are at all doubtful about, throw it away.

There are an enormous number of woodland toadstools. Some of these can only be found in certain kinds of woods. This is because they form symbiotic relationships with particular types of tree. The mycelium of the fungus becomes entangled with the roots of the tree to form a structure called a mycorrhiza. Both the fungus and the tree benefit from this relationship, especially in places where the soil is poor. The fly agaric (*see* AMANITA) forms mycorrhizas and is generally found in birchwoods.

Lichens

LICHENS are not single plants; each one is a symbiotic association of an alga and a fungus. The alga is green and makes its own food, some of which is used by the fungus. The alga benefits by getting protection from bright light and dry conditions. Lichens mostly contain single-celled green algae, such as *Pleurococcus*, or filamentous blue-green algae, such as *Nostoc*. The fungi involved are usually ASCOMYCETES.

Many different lichens occur. They may have a leafy or shrubby appearance, or they may form crusts on rocks. Some even grow in the surface layers of rocks. They take the minerals they need from the rainwater that flows over them. Lichens can withstand both extreme cold and heat, and are therefore among the first plant colonizers of regions where harsh conditions prevail. They help to break down the surface of the rocks on which they live, thus starting the process of soil formation.

Lichens can reproduce asexually by forming a powdery mass of structures that contain both the alga and the fungus. Sexual reproduction only involves the production of fungal spores, and a new lichen is formed only if the spores germinate near a suitable algal partner.

Above: Lichens are often found on heathland, where only small, hardy plants such as mosses can survive. The trumpet-shaped fruiting bodies of this lichen produce only fungal spores. A new lichen plant is formed when the spores germinate in the presence of a suitable alga. **Right:** Lichens may also be found on cliffs and rocks, where no other plants can grow.

---

cipuliforme is a similar fungus related to the GIANT PUFF BALL.

S **Saprophyte** is an organism that obtains its food as dissolved chemicals from decaying plant or animal tissue. All fungi are either saprophytes or PARASITES.

Stinkhorns are an order of basidiomycetes related to the PUFF BALLS. The spores of the stinkhorn (*Phallus impudicus*) are produced on the end of a short stalk inside the fruiting body. When they are ripe the stalk grows rapidly upwards, forcing its way through the outer covering. The spore mass gives off an unpleasant smell. This attracts flies,

Young shaggy ink caps

which swiftly disperse the spores.

Stipe is the stalk of a toadstool.

Symbiosis is an association between 2 living organisms that harms neither of them and may benefit one or both of them.

T **Toadstools** are the umbrella-shaped fruiting bodies of certain basidiomycetes. The name is commonly used to mean all those except the edible members of the genus AGARICUS, such as the field MUSHROOM.

Truffles (*Tuber*) are a genus

Stinkhorn

of ascomycetes, related to MORELS. They are difficult to find, but excellent to eat.

V **Veil** is the thin membrane that completely covers the young fruiting bodies of the genus AMANITA.

Y **Yeasts** include several types of single-celled ascomycetes, fungi that reproduce by budding. Many yeasts are capable of fermenting sugar to produce alcohol. The yeasts used in baking, brewing and wine making are *Saccharomyces cerevisiae*.

Mosses and liverworts are two related groups of mostly small, simple plants. They are great colonizers, because their tiny spores can be carried long distances to new land areas, where they germinate and produce new plants.

Mosses and Liverworts

Mosses and liverworts are simple, green plants that have no true roots. Instead they are attached to the ground by tiny root-like threads called rhizoids. They live on land and therefore have an advantage over the algae. But their environment must still be moist, because they have no means of preventing their cells from drying out a moist atmosphere is thus essential for the growth of mosses; it is also necessary for reproduction and for taking in oxygen. Dry conditions kill liverworts and many mosses. However, some mosses can survive even when water is scarce. They shrivel up and appear to be dead. When they are moistened again, they swell up and continue to grow.

Liverworts

The simplest liverworts have no stems or leaves. They have flat plant bodies that spread over the ground. Such plant bodies are called thalli (*see* THALLUS), and liverworts that have this kind of structure are called thalloid liverworts. PELLIA and MARCHANTIA are typical examples of thalloid liverworts.

A more advanced group are called leafy liverworts. These have distinct stems and thin, filmy leaves. However, their stems do not contain any water-conducting cells. The shape of the delicate leaves and the way in which they are arranged on the stem varies considerably. There are many species – over 90 per cent of all the liverworts are leafy liverworts. Examples include LOPHOCOLEA and CALYPOGEIA.

Mosses

Mosses are familiar plants that can be found growing in almost every environment. They are mostly small, low-growing plants with stems and leaves that form mats or small cushions. They only grow to a few centimetres high, but some species spread over a wide area. *Polytrichum*

commune is one of the largest European mosses, and may have stems 20 centimetres long. Species of *Dawsonia* in Australia may reach 70 centimetres. The largest British moss is *Fontinalis antipyretica*. It is common in freshwater streams and ponds, where it grows submerged under the water. It grows to a length of about 100 centimetres.

Many mosses, particularly those that can withstand dry conditions, live in places that are unsuitable for higher plants. Together with lichens, they can be found in the Arctic tundra, on rocky mountain tops, on walls, and on the bark of trees. Their ability to live in such inhospitable conditions makes them good pioneering plants. When a new island is formed, mosses and lichens are the first plants to colonize it. For example, in 1963, volcanic action caused

Right: *Polytrichum* is a genus of moss which grows on moorlands. At certain times of the year these mosses produce large numbers of orange-brown capsules, which contain the spores.

Below: Mosses can grow in places that other plants find inhospitable. These mosses are growing on Mount Kenya at 3,500 metres above sea level.

Reference

A **Annulus** (of a moss) is a ring of large cells round the top of a moss capsule. When the capsule is ripe, the cells of the annulus break and the top of the capsule falls off. The spores can then be released.
Archegonium. This is the female sex organ of a moss, liverwort or fern. It consists of a swollen base, or venter, and a long, hollow neck. The venter contains the female sex cell (oosphere).

B **Bog moss,** is the popular name for SPHAGNUM.
Bryophytes is the scientific name for mosses and liverworts. They make up the

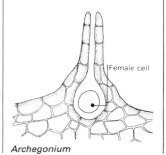
Archegonium

division Bryophyta of the plant kingdom.
Bryum is a genus of mosses. *B. capillare* is a common moss on walls and roofs, where it forms compact green cushions. *B. argenteum* also grows on walls and roofs, and in the cracks between paving stones. It is a dark green moss with a silvery sheen. This species has even been found in the Arizona desert. *B. pseudotriquetrum* grows near mountain streams.

C **Calypogeia** is a genus of leafy liverworts. *C.*

bicuspidata grows in peat bogs, on rotting timber, and on wet soil. *C. fissa* grows in peat bogs, on wet sandy soil, and on sandstone rocks. *C.*

Andreaea rupestris (moss)

meylanii grows on peaty and sandy soils, and on sandstone rocks.
Calyptra is the sheath that completely surrounds the capsule of a moss or liverwort as it develops. It is formed by the growth of the neck of the ARCHEGONIUM.
Capsule, see SPOROGONIUM.
Columella is the central column in a moss capsule. No spores are produced in this region of the capsule.

E **Elaters** are long, springy cells in the mature capsule of a liverwort. When the capsule opens, the ela-

Below: *Pellia epiphylla* is a thalloid liverwort found in moist, shady places. Its round, black capsules are produced in early spring.

Below: *Riccardia pinguis* is a small thalloid liverwort commonly found in wet places. Its lobes have no midribs.

Below: *Lophocolea cuspidata* is a leafy liverwort that grows on bark in damp woods. Its leaves have 2 large points at their tips.

Above: *Marchantia polymorpha* is a large thalloid liverwort found in marshy places.

Above: *Calypogeia meylanii* is a leafy liverwort that grows on peaty and sandy soils and on sandstone rocks.

ters help to flick the spores out.

Elaterophore. This is a mass of elaters attached to the inside of the capsule. When the capsule opens, some spores are trapped in the elaterophore and are released at a later time.

F **Fontinalis** is a genus of mosses. *F. antipyretica* is the largest of all the British mosses. It grows submerged underwater, attached to stones in slow-moving streams and rivers.

Funaria is a genus of mosses. *F. hygrometrica* is a common moss that forms patches in fields and on heaths. Its orange-brown capsules hang downwards on the ends of their stalks.

G **Gametophyte** is the generation during the life cycle of a plant that produces sex cells (gametes). The gametophytes of mosses and liverworts are the adult plants. The gametophyte of a fern is the PROTHALLUS (*see page 95*). Gymnosperms and flowering plants have separate male and female gametophytes. The male gametophyte is the pollen grain, and the female is found within the ovule.

H **Hepaticae** is the scientific name for the liver-

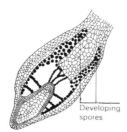

Developing spores

Moss capsule

worts; a class of the division Bryophyta.

Hypnum cupressiforme is a species of moss of which there are several varieties. Some grow on tree trunks, where they form silky tufts. One variety is found on chalk grassland; it has a rich green colour that may be tinged with yellow or bronze.

I **Involucre** is the outer protective covering of a moss or liverwort SPOROGONIUM. In thalloid liverworts it is formed from two flaps of the THALLUS, above and below the sporogonium. In leafy liverworts and mosses it is formed from leaves around the sporogonium.

L **Lophocolea** is a genus of leafy liverworts. *L bidentata* is a common liverwort found in grassy places, such as lawns. *L. cuspidata* is found on the bark of trees.

M **Marchantia** is a genus of thalloid liverworts. *M. polymorpha* has a large, spreading THALLUS. It is common in marshy places, the banks of streams, in gardens and greenhouses.

the island of Surtsey to emerge from the sea near Iceland. Within a few months spores blown from Iceland had landed on the bare rock, and the island was soon colonized by lichens and mosses. Their rhizoids helped to break down the surface of the rock. The powdered rock, together with the remains of dead mosses, began to form pockets of soil. Soon the first flowering plants were growing on the island.

However, the greatest number of mosses occur in damp places, and they grow best where there is little competition from higher plants. For example, the peat bogs are too acid for most plants. Thus the common plants in peat bogs are mosses that can tolerate this acidity. Large hummocks of *Sphagnum papillosum* form here, building up gradually over several years. Eventually the hummocks become so raised that other plants can grow on them. But generally the hummocks break up and the process begins again.

Mosses occur in many other damp places, such as woods and the banks of streams. Many of them will only grow on one type of soil, and it is possible to say whether the soil is acid (for example peat and sand), alkaline (chalk) or rich in nitrogen or phosphorus by identifying the mosses that grow on it.

How mosses and liverworts reproduce

The most important form of reproduction in the mosses and liverworts is sexual reproduction. In fact only the liverworts have any kind of asexual reproduction. *Marchantia* can produce small cup-shaped structures called gemmae on its leaves. When these fall off they can grow into new plants.

The sexual reproduction of mosses and liverworts is more complicated than that found in most algae. They have a method of increasing the number of new plants they produce. Large numbers are important because the young plants are unprotected, and many of them do not survive.

Mosses and liverworts have two generations. One generation produces gametes (sex cells). A fertilized female gamete becomes the second generation plant, which produces a mass of tiny spores. When these germinate, they grow into new gamete-producing plants. This process is

Below: *Mnium hornum* is a common woodland moss that forms large, dull green turfs on wood and peat. In spring the new shoots are a contrasting light green.

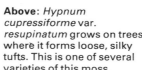

Above: *Hypnum cupressiforme* var. *resupinatum* grows on trees, where it forms loose, silky tufts. This is one of several varieties of this moss.

Below: *Bryum capillare* is a common moss on roofs and the tops of walls. Its spore capsules are green when young, but turn brown as they ripen.

Below: *Amphidium mougeotii* is a mountain moss that forms rounded cushions on wet rocks. The leaves of each plant are long and narrow.

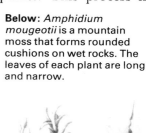

Above: *Sphagnum palustre* is a moss found in the drier, less acid parts of bogs, where it forms pale green mats. It has round spore capsules with circular lids.

Mnium is a genus of mosses. *M. hornum* is one of the most common mosses in woodlands, where it is often found growing on wood or peat. *M. punctatum* is a woodland moss found in shady places near streams. *M. undulatum* is light green moss that grows in shady places. It has upright stems that have a palm-like appearance.

Musci is the scientific name for the mosses; a class of the division Bryophyta.

O **Oosphere,** see ARCHEGONIUM.

Operculum is the cap of a moss capsule. When the capsule is ripe the ANNULUS breaks and the operculum falls off.

Leucobryum glaucum (moss)

P **Paraphyses** are many-celled hairs intermingled with the male and female sex organs (*antheridia* and *archegonia*) of a moss. They probably help to keep the sex organs moist.

Pellia is a genus of the thalloid liverworts. *P. epiphylla* grows on the banks of shady streams. *P. fabroniana* grows on wet limestone rocks or on chalky soils.

Peristome teeth are long pointed structures at the top of a moss capsule. They are exposed when the OPERCULUM falls off. They are arranged like the spokes of a wheel and in wet conditions they prevent spores from being released. In dry conditions, the teeth move farther apart, allowing spores to pass through the gaps. Thus spores are only shed in dry conditions, when there is a likelihood of air currents spreading them over a wide area.

Polytrichum is a genus of mosses. *P. commune* is the largest British land moss. It is common on wet, peaty moors. *P. formosum* also grows on moors, but is also found in woods and banks on acid soil. *P. piliferum* forms 'turfs' on sand dunes.

Protonema. This is the thread-like plant that is formed when a moss germi-

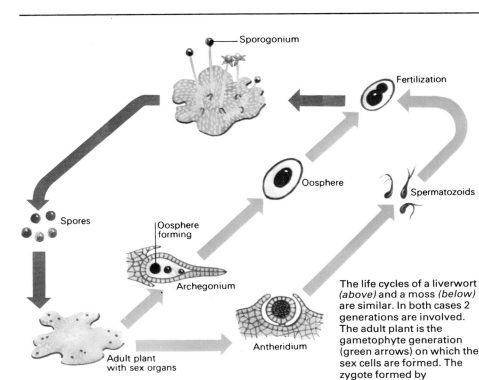

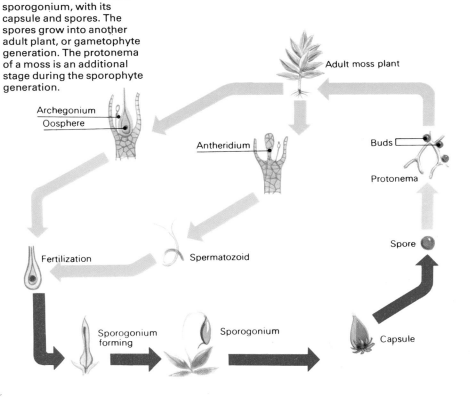

The life cycles of a liverwort *(above)* and a moss *(below)* are similar. In both cases 2 generations are involved. The adult plant is the gametophyte generation (green arrows) on which the sex cells are formed. The zygote formed by fertilization grows into the sporophyte generation (red arrows). This consists of a sporogonium, with its capsule and spores. The spores grow into another adult plant, or gametophyte generation. The protonema of a moss is an additional stage during the sporophyte generation.

called alternation of generations. The gamete-producing generation is called the GAMET-OPHYTE, and the spore-producing generation is called the SPOROPHYTE.

The adult plant of a liverwort is the gametophyte generation. On the upper surface of its THALLUS it grows male sex organs (antheridia), which produce male sex cells (SPER-MATOZOIDS). Round the edge of the thallus the plant grows female sex organs (ARCHEGONIA), each of which contains a single female sex cell (oosphere). When the sex organs are fully developed, the spermatozoids swim over the wet surface of the thallus, and one spermatozoid fertilizes each oosphere. The fertilized oosphere immediately begins to divide and becomes the second generation — the sporophyte.

The sporophyte generation consists of a capsule on a long stalk. This is called the SPOROGONIUM and it remains attached to the adult liverwort plant. Inside the capsule, meiosis occurs *(see page 75)* and spores are formed. The capsule eventually breaks open, scattering the spores. Each spore is capable of growing into a new liverwort plant and the cycle begins again.

The reproduction of a moss is very similar. The adult plant is the gametophyte, but in this case both male and female sex organs grow at the top of the stem. Again, a sporogonium is produced after fertilization, but its structure is more complicated. Mosses also add a further stage to their reproduction. Each spore can grow into a thread-like plant — rather like a filamentous alga — called a protonema. Each protonema produces several buds, which eventually grow into adult moss plants. Thus, by adding the protonema stage, mosses are again able to increase the number of possible new plants.

Alternation of generations is thus an efficient way of producing large numbers of new plants. However, in the sexual reproduction of the mosses and liverworts there is one great disadvantage. The adult plants are gametophytes (gamete-producing plants), and the male gametes (sex cells) need water in which to swim. Therefore, the adult plant must be wet before reproduction can occur. Higher plants have solved this problem, and therefore do not have to live in wet conditions.

---

nates. It produces several buds, each of which may grow into an adult plant.

Lunularia cruciata (liverwort)

R **Riccardia** is a genus of thalloid liverworts. *R.*

pinguis is one of the simplest liverworts; it grows in many wet places.

S **Sphagnum** is a genus of mosses, often referred to as bog mosses. They grow in dense masses, and their lower parts are continuously decaying slowly to form peat. There are several species, growing in varying degrees of wetness and acidity.

Spermatozoid is a single male sex cell with flagella, which it uses for swimming.

Spores are single- or many-celled structures formed during the reproductive processes of many plants. They are usually microscopic and are produced in large numbers.

Spore sac is the space inside the moss capsule that contains the spores.

Sporogonium is the structure that develops from a fertilized female sex cell of a moss or liverwort. It is the SPOROPHYTE generation and consists of a spore-containing capsule supported on a stalk, or seta. It is attached to the adult plant (GAMETOPHYTE) by a mass of cells called the foot.

Sporophyte. This is the generation during the life cycle of a plant that produces spores. The sporophyte of a moss or liverwort

Apple moss with capsules

is the SPOROGONIUM. The sporophytes of ferns, gymnosperms and flowering plants are the adult plants. Gymnosperms and flowering plants produce two kinds of spore. Microspores are pollen grains. Megaspores are single cells that remain inside the ovules.

T **Thallus** is a simple plant body that is not divided into root, stem and leaves.

Ferns, which may be as small as mosses or as large as trees, were common during the
Carboniferous period of the Earth's history. Their remains form a large part of the coal
seams which were formed at that time.

Ferns

Ferns, clubmosses, horsetails, quillworts and psilotes have two main things in common. Their adult plants produce spores, and their stems and leaves contain water-conducting cells. This gives them a great advantage over the mosses and liverworts. They do not have to live in wet conditions, and the adult plants can grow taller. Some of the clubmosses that existed during the Carboniferous period were large tree-like plants, and some modern tropical tree ferns can grow up to 25 metres high.

The first land plants that we know of were the psilophytes (*see page 68*), which became extinct about 370 million years ago during the Devonian period. Two modern plants, PSILOTUM and *Tmesipteris,* resemble them. These very simple plants are called psilotes, and it is possible that they are descended from the psilophytes, although we have no fossil evidence for this.

During the Carboniferous period there were a large number of lycophytes (clubmosses and quillworts). Today only five types exist. These include the clubmosses LYCOPODIUM and SELAGINELLA. Unlike some of their ancestors, such as the large tee-like *Lepidodendron* (*see page 68*), modern clubmosses are fairly small, creeping plants. The quillworts, which all belong to the genus ISOETES, are also descended from the lycophytes. These plants do not live on land. They have returned to the water and live permanently submerged. The horsetails are distantly related to the clubmosses and quillworts. In the Carboniferous period large horsetails existed, such as *Calamites,* but today species of EQUISETUM, the only genus that remains, are relatively small plants.

The ferns are a much larger group. There are about 10,000 species distributed all over the world. Most ferns are small plants, and some water ferns are less than one centimetre across. Only the tree ferns grow more than two metres

Below: The royal fern (*Osmunda regalis*) is the largest British fern. It grows in wet places, such as the banks of ponds and streams.

high. Tree ferns and many other types are only found in tropical and sub-tropical rain forests. In fact nearly 75 per cent of all the ferns live in these conditions. Only a few grow in cool and dry climates. In the tropical rain forests, smaller ferns are most often found living in the branches of trees. Here they receive more light than they would on the ground, which is often too swampy to support them. In cooler regions, ferns grow in the soil and among rocks. Most of them are adapted to living in damp and dimly-lit places, but some thrive in dry, sunny habitats and colonize open spaces.

A **Adder's tongue,** see OPHIOGLOSSUM.
Adiantum is a genus of ferns that grow in fairly dry regions. The maidenhair fern (*A. capillis-veris*) grows on sheltered limestone cliffs near the sea. Its delicate fronds have wedge-shaped leaflets.
Arthrophytes is the scientific name for the horsetails (*see* EQUISETUM): a subdivision of the division Pteridophyta.

Asplenium is a genus of ferns that grows on limestone walls or cliffs called spleenworts. Fronds vary according to species.

Ceterach officinarum

B **Botrychium** is a genus of 35 species of ferns found all over the world. *B. lunaria* (moonwort) is a common British fern that grows on rocky ledges and in grasslands. Each plant has a single frond that divides to form a fan-shaped, leafy part and a spherical spore-bearing part.
Bracken, see PTERIDIUM.

C **Clubmosses,** see LYCO-PODIUM, SELAGINELLA.

D **Dryopteris** is a genus of typical ferns with long fronds divided into 20-

35 main lobes. *D. filix-mas* (male fern), which grows in woodlands, is the commonest species in the British Isles.

E **Equisetum** is the only living genus of horse-tails. There are 20 species (10 of which are British), all of which grow in damp places. They have jointed stems, and at each joint there is a ring of small branches. There are no leaves. Spores are produced in cones (*see page 91*) at the tops of stems that are unbranched.

F **Filicophytes** is the scientific name for the ferns; a subdivision of the division Pteridophyta.

H **Hart's tongue,** see PHYLLITIS.
Horsetails, see EQUISETUM.

I **Indusium,** see SORUS.
Isoetes is a genus of water plants known as quillworts. They have a dense tuft of leaves that grow directly from a collection of thick, white roots. They live permanently submerged under the water in lakes and tarns.

Ferns have very small roots and their main underground organ is a creeping stem, or rhizome. This grows horizontally through the ground, putting up new shoots at intervals. In this way ferns can spread over a wide area. The leaves of a young shoot are tightly curled, and they unfold gradually as they grow. Some ferns have very simple, strap-shaped leaves. Others have complicated leaves that are divided many times. These leaves are called fronds, and the smaller divisions are called pinnae.

A few ferns are adapted to special ways of life. For example *Azolla* and *Salvinia* live in water. Their leaves float on the surface and their roots hang down. The upper surface of the fronds is covered with specialized hairs that repel water and ensure that the plant does not sink. Other ferns have methods of overcoming dry conditions. Some species of ASPLENIUM have leaves that shrivel when there is a lack of water, and a number of ferns have fleshy leaves that store water. Ferns that live in the branches of trees have a particular problem in getting all the food

Below: *Cyathea smithii* is a tree fern found in New Zealand. Tree ferns are common throughout the tropics.
Below right: Some fern relatives. *Equisetum telmateia* is a horsetail. The stag's horn moss, *Lycopodium clavatum* is a mountain clubmoss. *Psilotum nudum* is a psilote found throughout the tropics and subtropics. *Isoetes lacustris* is a quillwort that grows underwater in lakes.

they need. Species of *Platycerium,* a type of tropical fern, have overcome this problem. Each plant has special fronds near the base that trap dead plant material. The plant can then absorb what it needs from the resulting humus that forms.

How ferns reproduce
Ferns reproduce asexually when their spreading rhizomes put up new shoots. In swampy tropical rain forests, some ferns produce small plantlets on their fronds. These drop off and grow into new plants. In such conditions this is often the most important form of reproduction.

The sexual reproduction of ferns, like that of the mosses and liverworts, involves the alternation of two generations. In this case the adult plant is the sporophyte (spore-producing plant) and the gametophyte (gamete-producing plant) is very small. In addition, the two plant generations are separate.

Spores are produced in structures called SPORANGIA on the undersides of certain fronds.

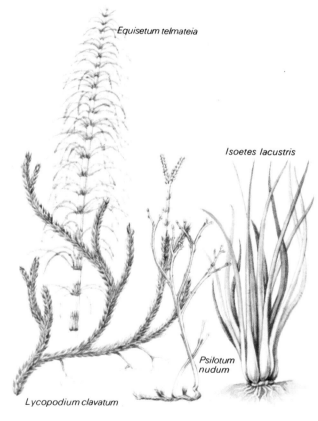

Equisetum telmateia

Isoetes lacustris

Psilotum nudum

Lycopodium clavatum

L **Lycophytes** is the scientific name for the clubmosses (see LYCOPODIUM, SELAGINELLA) and quillworts (see ISOETES); a subdivision

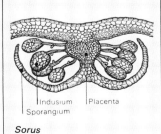

Indusium Placenta
Sporangium
Sorus

of the division Pteridophyta. **Lycopodium** is a genus of clubmosses. There are about 200 species, most of which occur in tropical regions. Some are found in alpine areas and in the Arctic. There are 4 British species. They have long slender stems covered in small leaves.

M **Maidenhair fern,** see ADIANTUM.
Marsilea is a genus of water ferns. There are 65 species and they occur in both tropical and temperate regions. The fronds grow from an

underwater creeping rhizome, and each frond bears 4 leaflets, rather like a 4-leaf clover.

O **Ophioglossum** is a genus of 45 species of ferns found in almost all the countries of the world. *O. vulgatum* (adder's tongue) is common in grassland in Britain. Like BOTRYCHIUM, to which it is related, it has a single frond. This divides to form an oval leaf and a spore-bearing spike.
Osmunda is a genus of 14 species of ferns found all over the world. *O. regalis*

(royal fern) is the largest British fern. It grows in wet places, and the leaflets on the fronds are undivided.

P **Phyllitis** is a genus of ferns related to AS-PLENIUM. *P. scolopendrium* (hart's tongue) is the only British fern that has completely undivided, strap-shaped fronds.
Prothallus. A small, green, flat plant, often heart-shaped, that is the gametophyte (see page 91) generation of a fern.
Psilotum is a genus of 2 species of primitive plants.

They occur in tropical and sub-tropical regions and grow in the ground among rocks and also in the branches of trees.

Jamaican tree fern

Below: The life cycle of a fern involves 2 separate generations. The adult plant is the spore producing plant, or sporophyte (red arrows). A single spore develops into a prothallus, or gametophyte – the plant that produces sex cells (green arrows). After fertilization the zygote develops into a new adult plant.

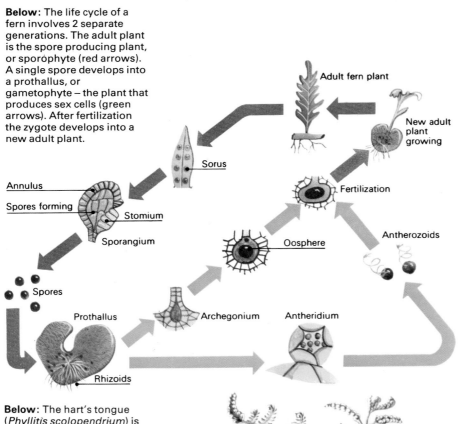

Several sporangia are collected together in a group called a SORUS. You can see these as brown dots on the underneath of a spore-producing frond.

When the sporangia are ripe, the spores are released into the air. They germinate on the ground, and each spore may grow into a tiny heart-shaped plant called a PROTHALLUS. This is the gametophyte generation. Male and female sex organs develop on the upper surface of the prothallus.

The male organs (antheridia) produce male sex cells (antherozoids). These swim to the female organs (archegonia) and fertilize the female sex cells (oospheres). One fertilized oosphere on each prothallus then develops into a new adult plant.

As in the reproduction of the mosses and liverworts, alternation of generations in the fern life cycle ensures that large numbers of new plants are produced. However, in this case the adult plant is the sporophyte, and therefore does not have to live in wet conditions. Only the small gametophyte has to be moist, so that the male sex cells can swim to the female organs.

Below: The hart's tongue (*Phyllitis scolopendrium*) is the only British fern that has undivided, strap-like fronds.

Above: Bracken (*Pteridium aquilinum*) is the most widespread of all ferns, being found in the tropics as well as in temperate regions.

Above: The prickly shield fern (*Polystichum aculeatum*) is a common fern in woods and hedges throughout Britain.

Above: Moonwort (*Botrychium lunaria*) is a fern that grows in grassland and ledges on mountainsides.

Pteridium is a genus of ferns that can successfully compete with flowering plants. This may be due to their very deep systems of rhizomes. *P. aquilinum* (bracken) is the most widespread of all ferns, being found in the tropics, in temperate regions, and even near the Arctic Circle.

Pteridophytes are a group of plants that include the ferns and the 'fern allies' – clubmosses, horsetails, quillworts and psilotes.

Q **Quillworts,** see ISOETES.

S **Selaginella** is a genus of more than 700 species of clubmosses. Most of these occur in tropical and sub-tropical regions, but some grow in temperate areas. They are generally found in damp places, such as the floor of rain forests. Some can survive desert conditions. They are similar to LYCOPODIUM in form.

Sorus (*plural*: sori) is a group of fern sporangia on the underside of a frond. The stalks of the sporangia are attached to a central structure called the placenta. The whole sorus is often covered by a flap called the indusium.

Spleenwort, see ASPLENIUM.

Sporangium. This is the spore-containing structure

Epiphytic fern – a fern that lives on a tree

of a fern. It consists of a capsule on the end of a stalk. A special row of cells runs over the capsule. Some of these cells have thick walls and form the annulus. The remainder have thin walls and form the stomium. When the sporangium is ripe, the cells of the annulus stretch, and the capsule breaks open at the stomium, releasing the spores.

T **Tmesipteris** is a genus of 2 species of primitive plants found only in Australasia. They grow both in the ground and in the branches of trees. They are related to PSILOTUM, but have larger leaves.

The term gymnosperm comes from two Greek words, meaning 'naked seed'. The most familiar gymnosperms are the conifers, whose seeds are contained in cones. Conifers include some of the world's largest and longest-living trees.

Gymnosperms

Gymnosperms are seed-bearing plants. But, unlike the flowering plants (angiosperms), their seeds are partly exposed. For example, the seeds inside the cone of a pine tree are only protected by scales, which can be lifted off. The seeds of a flowering plant on the other hand, are completely surrounded by the wall of the ovary (*see page 104*). The word gymnosperm means 'naked seed'; angiosperm means 'enclosed seed'.

The best-known gymnosperms are the conifers, which all bear their seeds in cones. But YEWS, CYCADS and Gnetales are also gymnosperms. And the MAIDENHAIR TREE is the only living member of a group of gymnosperms that flourished in the Mesozoic era.

Below: *Cycas media* is a cycad found in Queensland, Australia. There are only 15 species of cycads living.

Above: *Welwitschia* is a strange gymnosperm with long, trailing leaves and cones borne on a short stem.

Yews are evergreen trees found in Europe, Asia and North America. They have long narrow leaves, and for this reason they are often called conifers. However, they do not bear their seeds in cones. Each seed is borne separately and is almost completely enclosed by a fleshy, red, cup-shaped structure that resembles a berry.

Cycads are palm-like trees found only in tropical areas. They have separate male and female plants, and their seeds are produced in large cones in the centre of the female plants.

The Gnetales are a strange group of plants that have some of the features of flowering plants. For example, the water-conducting vessels in their stems are the same as those of flowering plants. However, they bear naked seeds and are therefore generally classed as gymnosperms. The three genera, GNETUM, EPHEDRA and WELWITSCHIA are totally unalike.

The true conifers are the largest group of gymnosperms. They include the PINES, SPRUCES, FIRS, CEDARS, LARCHES and CYPRESSES. They are all evergreen except for the larches and the swamp cypress. They have long, needle-shaped leaves. Each needle is tough and leathery and has a

Reference

C **Cedars** (*Cedrus*) are a genus of 4 species of conifers. The three most important species are the Atlas cedar (*C. atlantica*), which grows in the Atlas mountains of North Africa; the cedar of Lebanon (*C. libani*) from south-western Asia; and the deodar (*C. deodara*) from the western Himalayas.
Cycads (*Cycas*) are primitive gymnosperms found only in the tropics. They have large feathery leaves

that arise from an un-branched stem. The lower part of the stem has leaf scars.
Cypresses (*Cupressus*) are a genus of 20 species of

Cedar of Lebanon

trees found in North America, Europe and parts of Asia. They have small, scale-like leaves and round cones.

D **Dawn redwood** (*Metasequoia glyptostroboides*) is a deciduous conifer that grows up to 35 metres high. Until 1944 it was only known from fossil remains, and was thought to be extinct. But a small number were found in south-western China, and the tree is now cultivated in gardens as an ornamental.
Douglas fir (*Pseudotsuga menziesii*). A tall conifer (up

to 90 metres) found in western America. It has reddish-yellow timber that is used in many kinds of construction, and its tall trunks are used for masts and poles.

E **Ephedra** is a genus of 40 species, most of which are shrubs. They are found in warm deserts in North and South America, and in a belt across Asia from the Mediterranean to China.
Evergreen trees and shrubs are those that have green leaves at all times of the year. Most conifers are evergreen because their

leaves do not fall all at once, but over a 3-year period.

F **Firs** (*Abies*) are a genus of tall, pyramid-shaped

Cone of Lebanon cedar

Above: There is a wide variety of conifers, which differ in the shape of the tree and its leaves. Those shown are *(left to right)* giant redwood *(Sequoiadendron giganteum)*, cypress *(Cupressus)*, Scots pine *(Pinus sylvestris)*, larch *(Larix)*, and Douglas fir *(Pseudotsuga menziesii)*.

concave surface inside and a convex surface outside. In hot, dry weather the leaf contracts so that it becomes almost cylindrical and this shape helps to reduce the rate of transpiration. As a result conifers can grow in dry places, and also in cold places, where the amount of water in the soil is too low for broad-leaved trees. They can also grow in poor soils that contain few minerals.

Most conifers are found in the Northern Hemisphere. A broad band of coniferous forest stretches round the world just below the tundra zone, and some forests grow as far north as the Arctic Circle. Some conifers are found in hot, dry climates, such as the Mediterranean region and on tropical mountainsides. But only a few occur in the Southern Hemisphere, such as the monkey

puzzle tree in the Andes of South America.

Conifers grow quickly and therefore they are useful timber trees. They are called softwoods because their wood is softer and easier to work than broad-leaved trees (hardwoods). Softwoods contain long fibres, and this makes them particularly useful for making paper. They are also used for making plywood, chipboard (chips of wood stuck together with plastic glue), furniture, house-building timber, telegraph poles, mine props, and fence posts.

How a pine tree reproduces

A pine tree, like many other gymnosperms, produces separate male and female reproductive organs on the same tree. The reproductive

conifers, with needle-like leaves. There are 35 species. The European silver fir (*A. alba*) is an important timber tree, and the American balsam fir (*A. balsamea*) is a source of turpentine balsam.

G Giant redwood (*Sequoiadendron giganteum*) is the most massive tree in the world, also called giant sequoia, California big tree and Wellingtonia. It is a conifer that grows to over 80 metres high. The largest tree, named General Sherman, is in the Sequoia National Park in California. It is

83 metres tall and has a girth of over 24 metres.

Gnetum is a genus of 40 species of plants that grow in tropical rain forests. They

Maritime pine

are mostly climbing plants, but a few are trees and shrubs. They produce their seeds in cones, but their leaves are identical to those of flowering plants.

H Hemlock (*Tsuga*) is a genus of 10 species of conifers. They are pyramidal in shape and their leaves are needle-like. The western hemlock (*T. heterophylla*) of North America grows to 60 metres high, and its timber is used for paper-making and construction. These trees should not be confused with the herbace-

ous hemlocks, which are poisonous flowering plants.

J Junipers (*Juniperus*) are a genus of 60 species

Juniper berries

of trees and shrubs found all over the Northern Hemisphere. They are conifers and the male and female cones develop on separate trees. The female cones develop in an unusual way, resembling bluish berries. The 'berries' of the common juniper (*J. communis*) are used for flavouring gin.

L Larches (*Larix*) are a genus of 10 species of deciduous conifers. The European larch (*L. decidua*) is found in mountainous areas and is often planted in gardens.

Left: The maidenhair tree (*Ginkgo biloba*) is a 'living fossil'. It is the only living member of a group that flourished 200 million years ago. It gets its Latin name from its 2-lobed leaf (*left*). It is native to China but is rare as a wild tree. It grows well in temperate regions and is often planted in gardens.

fertilize the egg cells.' However, only one fertilized egg cell in each ovule develops any further. It begins to divide and grows into an embryo. The cells that surround the egg cell develop into the other parts of the seed. Eventually, each scale in the female cone carries two seeds side by side. Six months after fertilization, in the following spring, the scales of the female cone open again and the seeds are released. Each one has a wing to enable it to be dispersed by the wind.

organs of a pine tree are the cones. Male cones develop from buds in the spring, and several cones may grow from a single bud. Each cone consists of a central rod-like structure surrounded by a number of scales. During the year each scale develops a pollen sac, and by the following spring the pollen is ready to be released. The scales move slightly apart and the pollen sac breaks open. The pollen grains each have two air sacs and they are carried by the wind to the female cones.

The first stage in the development of a female cone also takes about a year. The general structure is the same as that of a male cone. But instead of a pollen sac, each scale produces two female organs called ovules. Pollination occurs in the spring. The scales of the female cone open slightly, allowing the pollen to enter. The ovules each produce a tiny drop of sticky liquid in which the pollen grains are trapped. The scales of the female cone close up again, but fertilization does not occur until the summer of the following year – 18 months later. During this time two egg cells form inside each ovule. Then the sticky liquid dries up, and the pollen grains are released into the ovules.

Special nuclei from the pollen grains then

Right: The life cycle of a pine tree. Each tree produces separate male and female cones. The male cones produce winged pollen grains. The female cones produce 2 ovules on each scale, and each ovule develops 2 archegonia (female sex organs). Fertilization occurs when a nucleus from a pollen grain fuses with the oosphere in an archegonium. Only one archegonium in each ovule develops into an embryo, and the surrounding tissues form the seed coat. When the seeds are fully developed, each scale of the female cone has 2 winged seeds. These are dispersed by the wind, and in favourable conditions an embryo develops into a pine seedling.

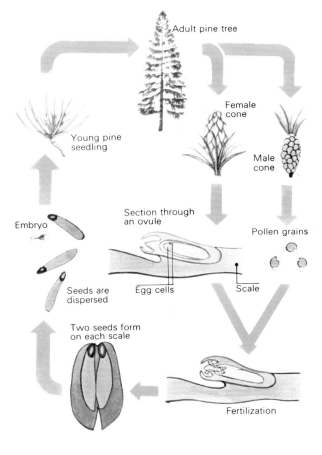

Adult pine tree

Young pine seedling

Female cone

Male cone

Section through an ovule

Embryo

Pollen grains

Seeds are dispersed

Egg cells

Scale

Two seeds form on each scale

Fertilization

Flowering plants, which dominate our countryside and gardens, form the most varied of all plant groups. Botanists call them angiosperms, the Greek for 'enclosed seeds'. They form about 75 per cent of all land-dwelling plants.

Flowering Plants

Flowering plants dominate the plant world. Over the last 65 million years they have become so successful that there are now only a few places in which they cannot be found. Their success is due partly to the fact that they have been able to adapt to many different environments. As a result, they are not only a very large group, but also an incredibly varied group. Such very different plants as grasses, trees and cacti are all flowering plants, as well as the more obvious flowers of the countryside.

The flowering plants are divided into two main classes. Monocotyledons (sometimes shortened to monocots) are all those plants that have long thin leaves with veins that run parallel to each

Below right: Flowering plants with long, thin leaves and 1 seed leaf are called monocotyledons
Below: Flowering plants range from very simple blooms to intricate arrangements such as in this bird of paradise flower (*Strelitzia regina*).

other. They get their name from the fact that their seeds contain only one seed leaf, or COTYLEDON. Most monocotyledons in temperate climates are small, non-woody plants. But some tropical monocotyledons, such as PALM trees, are large. Dicotyledons (or dicots) are more varied, but basically their leaves are broader and have a network of veins. Their seeds have two cotyledons.

A flowering plant consists of a root system, stem, leaves, and one or more flowers. The root systems vary considerably. A typical system consists of a single primary root that has branches, or secondary roots. The secondary roots themselves branch repeatedly. The smallest

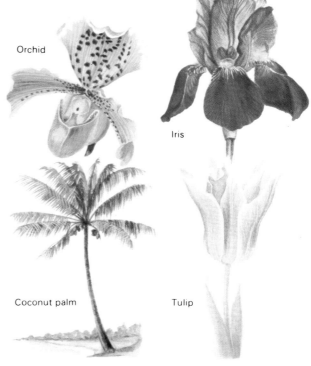
Orchid
Iris
Coconut palm
Tulip

Reference

A **Acacia** is a genus of about 500 species of thorny shrubs, mostly found in Australia. They are members of the PEA FAMILY and most species have bright yellow flowers. Their seeds are contained in pods.
Achenes are dry, one-seeded INDEHISCENT FRUITS; e.g. the fruits of the buttercup and dandelion.
Alder (*Alnus glutinosa*). A small British deciduous tree that has dark green leaves and bears catkins. Other alders are found all over the Northern Hemisphere.
Amaryllis family, includes the belladonna lily (*Amaryllis belladonna*), which is a South African plant with bright red flowers. Also in this family are SNOWDROPS, DAFFODILS and the AMERICAN CENTURY PLANT (*see page 113*).
Anemone is a genus of HERBACEOUS plants. A number of species are found all over the world; e.g. the wood anemone in Britain, which produces carpets of pinkish-white flowers in spring.
Annuals are plants that grow, produce seeds, and die within a single season.

Acacia

Antirrhinum is a genus of plants, commonly called snapdragons. They have dense spikes of red or purple flowers and are found all over the Northern Hemisphere.
Ash (*Fraxinus*) is a genus of about 60 species of deciduous trees found in the cold and temperate regions of the Northern Hemisphere. The common ash (*F. excelsior*) of Europe and Turkey grows to over 30 metres tall.

B **Beech** (*Fagus*) includes 10 species of deciduous tree that occur throughout the Northern Hemisphere. They are useful timber trees and the most important are the European beech (*F. sylvatica*) and the North Ameri-

Flowering ash

Above: Wheat belongs to the grass family, which is the largest of all the flowering plant families.
Below: Dictotyledons have broad leaves and 2 seed leaves.

roots are very fine and bear the root hairs that absorb water from the soil. GRASSES have no primary roots. Their root systems are just masses of branching threads. Some plants, such as thistles and dandelions, have long, thick, primary roots called tap roots. These penetrate deep into the soil, seeking out the water that other plants cannot reach.

The stem of a flowering plant is the supporting structure for the leaves and flowers. At its tip is a terminal bud, which is the main growing point of the stem. A simple stem bears leaves at points called nodes. Each leaf is attached to the stem by a leaf stalk, or petiole. Where the petiole joins the stem, an axillary bud forms. This may develop into a side branch. When this happens, the original leaf falls off, leaving a leaf scar, and the branch continues to grow. In woody plants the leaf scar eventually disappears as bark forms and the stem and branch increase in size.

Flowers
When a plant reaches a particular stage in its development it produces one or more flowers. These are shoots specially designed for reproduc-

tion. A flower may grow in place of a leaf, or from a terminal or axillary bud.

The main supporting part of a flower is called the receptacle. The remainder of the flower consists of four different sets of organs attached to the receptacle. The outer ring is made up of leaf-like structures called sepals, which are usually green. Inside these are the petals. Insects are attracted to certain colours, and therefore insect-pollinated flowers usually have brightly coloured petals.

The last two sets of organs are the reproductive parts. The male organs (stamens) consist of pollen sacs (anthers) on the ends of long filaments. In the centre of the flower are one or more female organs (carpels). The main part of a carpel is the ovary. This may contain one or more ovules, which contain the egg cells. Projecting from the top of the ovary is a long stalk, called the style. This bears a flat pollen-receiving surface, called the stigma, at the top.

Flowers may be produced singly, as in the case of the tulip. But they are often arranged in a group, or inflorescence. In many inflorescences the individual flowers can clearly be seen. They

Violet

Dog rose

Buttercup

Sunflower

White willow

Oak

Rhododendron

can beech (*F. grandifolia*). The popular copper beech is a variety of the European beech.
Bee orchid (*Ophrys apifera*) is a rare orchid whose flower has a broad velvety lip that resembles a bee. It grows on chalkland.
Berries are fruits that have fleshy or pulpy PERICARPS enclosing the seed. They include bananas, gooseberries, marrows, oranges, and dates. *See also* DRUPE.
Biennials are plants that grow, produce seeds, and die within a period of 2 years. During the first year

they store up food to be used in the second year, when they produce flowers and seeds. *See also* ANNUALS, PERENNIALS.

Lesser bindweed

Bindweeds are climbing plants that climb by twining their stems anti-clockwise round the stems of other plants. The larger bindweed (*Calystega sepium*) has trumpet-shaped white flowers. The small bindweed (*Convolvulus arvensis*) has cone-shaped pink or white flowers.
Birch (*Betula*) is a genus of about 40 species of deciduous tree found all over the Northern Hemisphere. The bark of these trees is smooth and peels off in layers. It is often silvery-white in colour, like that of the silver birch (*B.*

pendula). Other species have black, brown, red, orange or yellowish bark.
Bluebell, see HYACINTH.
Buttercups are a group of flowering plants belonging to the genus *Ranunculus*. This genus also includes the crowfoots and spearworts. Buttercups have cup-shaped, yellow flowers.

C Calyx is the outer ring of sepals of a flower. They are usually green, but in some flowers sepals and petals are the same colour.
Campions are members of the PINK FAMILY. They include

Red clover

the white campion (*Silene alba*) and the red campion (*S. dioica*), both of which are common in Britain.
Capsule is a dry, DEHISCENT

Horse chestnut stem

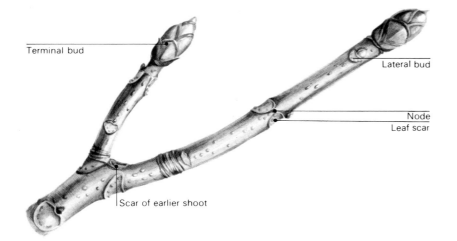

Terminal bud

Lateral bud

Node
Leaf scar

Scar of earlier shoot

Field geranium

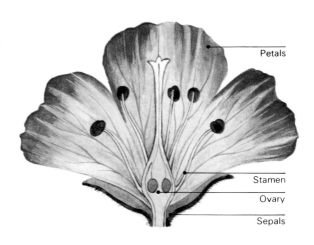

Petals

Stamen

Ovary

Sepals

Above: A horse chestnut shoot showing the main parts of a plant stem. In spring the terminal buds open to produce leaves and flowers.

Above right: A section through the flower of a field geranium. The male parts, or stamens, consist of the anthers and filaments. The female part, or carpel, consists of the stigma, style and ovary.

are produced in this way by a branching of the original reproductive shoot. The tall flowering shoot of the FOXGLOVE is an example of the simplest type of inflorescence. However, in some cases the inflorescence becomes so condensed that it looks like a single flower. The 'flower' of a dandelion is in fact made up of a number of tiny individual flowers, or florets.

Pollination

The next stage in the reproduction of a flowering plant involves the transfer of pollen from the anthers of one flower to the stigma of another. This is called cross-pollination. The transfer of pollen from the anthers to the stigma of the same flower is called self-pollination. But this is less desirable. Cross-pollination tends to produce stronger and healthier plants. Self-pollination does occur, but this is usually an accident. Many flowers have evolved ingenious ways of preventing self-pollination.

Water, wind, and animals are the three main agents for transferring pollen. Pollination by water is rare, and most pollens are damaged by water. In tropical countries some plants are pollinated by hummingbirds, bats, and sometimes even mammals but the vast majority of plants are pollinated by insects or wind.

Wind-pollinated plants produce light, dusty pollen. Large amounts of this pollen are needed to ensure that some of it reaches other plants.

Grasses are typical examples of wind-pollinated plants. A grass flower spills out pollen from large anthers on the ends of long, hanging filaments. Some of this pollen is collected by the feathery stigmas of other grass plants. Maize produces two kinds of flower on each plant, but the male and female flowers are well separated. This ensures that the pollen has a greater chance of reaching the female flowers on other nearby plants. Alder and hazel trees, too, have separate male and female flowers, which grow in inflorescences called catkins. WILLOW trees also produce catkins, but in this case the male and female catkins are grown on completely separate trees.

Insect-pollinated plants first have to attract insects to their flowers. Here, the colours of flowers play a large part. Experiments have shown that insects respond particularly to blues, mauves and purples. They see red only as a shade of grey. Thus in temperate climates there are very few naturally occurring pure red flowers. Some insects also respond to varying shades of ultra-violet light, which is invisible to humans. Thus some flowers that seem white to us do not appear white to insects.

In addition to colour, flowers often have particular markings, such as lines and coloured spots. These help the insects to find their way into flowers. The bee orchid has an even more elaborate arrangement. The flowers resemble female bees in appearance, odour and feel and

FRUIT that usually contains many seeds. The fruits of poppies, ANTIRRHINUMS, and irises are capsules.
Clematis is a genus of climbing plants that includes old man's beard (*C. vitalba*). This gets its name from the masses of feathery fruits produced by each flower. It is also called traveller's joy.
Clover (*Trifolium*) is a genus of over 300 species of HERBACEOUS plants. They are found all over the Northern Hemisphere and South America. The majority of them have 3 leaves joined at the base. Their flowers are

small, arranged in a round head and may be red, white or yellow.
Corolla is the petals of a flower.

Crocus

Cotyledons are the seed leaves that form part of the embryo in a seed.
Cow parsley (*Anthriscus sylvestris*). A tall fern-like plant that bears masses of white flowers in flattened heads. It is a very common plant.
Cowslip (*Primula veris*). A herbaceous plant related to the PRIMROSE. However, its flowers and leaves are smaller than the primrose's.
Crocus is a genus of monocotyledons consisting of about 80 species. The petals and sepals of their flowers are alike.

D **Daffodil** (*Narcissus*) is a genus of about 50 species of monocotyledons found in Europe, Asia, and North Africa. The flowers

Ox-eye daisy

have 6 white or yellow outer segments. Inside these there is a trumpet-shaped tube, which is generally yellow, orange or red.
Daisy (*Bellis*) is a genus of small HERBACEOUS plants found in many parts of the world. *B. perennis* is a common weed on lawns in Britain. The flower head of a daisy is made up of yellow disk florets surrounded by a ring of white ray florets.
Dandelion (*Taraxacum*) is a genus of herbaceous plants related to the daisy. Their flower heads are made up of yellow ray florets.

thus attract the male bees, who pollinate the flowers.

Some flowers attract bees, others attract moths, butterflies, beetles, or other insects. An insect attracted to the flower, must then be stimulated so that it performs the right actions. The nectar of a flower is the usual stimulant. It is produced in glands called nectaries at the bases of the petals. In drinking the nectar an insect picks up pollen from the anthers. At the same time it may brush pollen it is carrying from another flower on to the stigma.

Like wind-pollinated flowers, insect-pollinated varieties have various ways of preventing self-pollination, or at least reducing the possibility of its occurrence. For example, in some flowers the stamens and stigmas ripen at different times.

Other plants may produce two or three different types of flower. For example, the PRIMROSE has pin-eyed and thrum-eyed flowers on separate plants. The pin-eyed flowers have stamens well below the prominent stigma. In thrum-eyed flowers the positions are reversed, and the stamens are above the stigma. An insect visiting a pin-eyed flower gathers pollen on the

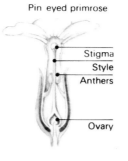
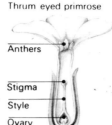

Pin eyed primrose

Stigma
Style
Anthers

Ovary

Thrum eyed primrose

Anthers

Stigma
Style
Ovary

front end of its body. When it visits a thrum-eyed flower, the pollen is brushed on to the stigma, and the insect collects pollen on the rear end of its body. This pollen can then pollinate a pin-eyed flower. The purple LOOSESTRIFE has an even more complicated arrangement of three types of flower. Pollen from each type can be used to pollinate the other two.

Many flowers are shaped so that an insect can only enter in a certain way. These flowers often have elaborate mechanisms for preventing self-pollination. In the violet flower, the four lower petals are arranged as a landing platform for the insect. Inside the flower the pollen-receiving part of the stigma is covered by a flap. A visiting insect automatically pushes open this flap and deposits pollen on to the stigma. The construction of the anthers is also modified so that a kind of pollen-collecting box is formed. The insect is showered with pollen from the box as it pushes into the flower to drink the nectar. This pollen cannot be deposited on the stigma because the flap closes as the insect backs out of the flower.

A few flowers use scent rather than colour to attract the insects in the first place. The LORDS-

Above: Primroses have 2 kinds of flower in order to prevent self-pollination. The anthers and stigmas are in different positions. Pollen from one kind of flower is transferred to the stigma of the other kind.

Left: Catkins of the silver birch tree shedding pollen into the wind.

Right: The flowers of bee orchids look like female bees. This attracts male bees, which pollinate the flowers.

Deadly nightshade, see NIGHTSHADE FAMILY.
Deadnettle, see NETTLE.
Deciduous tree. A tree that sheds its leaves at a particular time of year, usually the autumn. *See also* EVERGREEN.
Dehiscent fruit. A fruit that opens in an organized manner in order to release its seeds. *See also* INDEHISCENT FRUIT.
Dioecious plant. A species of plant in which male and female flowers are borne on separate individuals. *See also* MONOECIOUS.
Dogwood (Cornus) is a genus of about 40 species of

deciduous trees, found in northern temperate regions. The European red dogwood (*C. sanguinea*) has leaves that turn blood-red in

Deadly nightshade

autumn, and its bark turns red in winter.
Drupe. A fruit in which the outer layers of the PERICARP are fleshy or fibrous. But, unlike a BERRY, the inner layers surrounding the seed are stony. Examples of drupes include plums, cherries, peaches and coconuts. Blackberries and similar fruits are collections of drupes.

E **Elder** (Sambucus nigra) is a European shrub that grows to about 10 metres high. Although the shrub smells unpleasant, its

Elder

cream-coloured flowers and black fruits can be used to make excellent wine.
Elm (Ulmus) is a handsome deciduous tree that can

grow to over 30 metres tall. There are about 20 species found all over the Northern Hemisphere. However the English elm (*U. procera*), one of the grandest species, is being severely reduced in numbers by Dutch elm disease – a fungal disease carried by beetles.
Epigeal germination. A form of seed germination in which the cotyledons emerge from the seed case and photosynthesize. *See also* HYPOGEAL GERMINATION.
Endospermic seed. A seed in which the endosperm (food store) lies outside the

Left: A lords-and-ladies flower head cut away to show the flowers inside the green sheath. Small insects fall into the bottom of the sheath, where they pollinate the female flowers.

From flower to seed

After a pollen grain has landed on a stigma, it begins to develop a tube. This grows down through the style. Inside the tube are two sperm nuclei and a tube nucleus. When the tube reaches the ovule, fertilization occurs. One sperm nucleus fuses with the ovum, which begins to develop into an embryo. The other sperm nucleus fuses with the two cells next to the ovum, called polar cells. The resulting cell grows into the seed food-store, or endosperm.

Fruits and seeds

A mature seed consists of the embryo and its endosperm surrounded by a seed coat, or testa. The testa is formed from the layers that once surrounded the ovule. The seed is contained in a fruit, which is formed from the ovary wall.

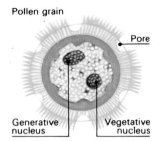

Pollen grain

Pore

Generative nucleus

Vegetative nucleus

AND-LADIES flowering shoot consists of a spike of flowers surrounded by a green sheath. The flowers are arranged on the spike in two groups – female near the base and male above. Above the flowers is a fringe of hairs and the whole spike is covered in a sheath. At the top of the spike is a club-shaped organ that produces a rotten smell, which attracts insects. The inside of the green sheath is slippery and small insects slide down and become trapped. Then they are held during the night. The downward pointing hairs prevent large insects from entering. If the small insects are carrying pollen from another flower-head, the female flowers will be pollinated. Then the male flowers ripen and shower the insects with pollen. By next morning the slippery surface has disintegrated, thus allowing the insects to escape carrying the pollen. Sooner or later the process will be repeated.

Right and above right: A pollen grain contains a generative nucleus and a vegetative nucleus. When it lands on a stigma the generative nucleus divides into 2 sperm nuclei. The pollen grain then grows a tube, which pushes its way down the style. When it reaches the ovule, the vegetative nucleus disintegrates and one sperm nucleus fuses with the ovum. This then grows into the embryo. The other sperm nucleus fuses with some of the nuclei surrounding the ovum. This then grows into the endosperm (food store) of the seed.

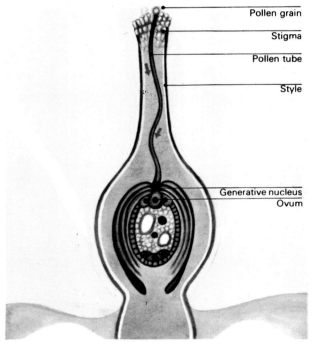

Pollen grain

Stigma

Pollen tube

Style

Generative nucleus
Ovum

COTYLEDONS. As the embryo grows, the cotyledons absorb the endosperm.
Eucalyptus is a genus of over 500 species of shrubs

Foxgloves

and trees, found in Australia and Tasmania. Many species grow to over 30 metres high, and the giant gum of Victoria (*E. regnans*) may reach 90 metres. Some species are grown for the oil that is produced in their leaves.
Evergreen is a shrub or tree that bears leaves all year. *See also* DECIDUOUS.

F Follicle is a dry fruit, formed from a single carpel, that opens by splitting down one side of the fruit. The fruits of monkshood (*Aconitum*) are follicles.

Forget-me-not (*Myosotis*) is a genus of HERBACEOUS plants with small, blue flowers. They are common in Europe, except for the mountain forget-me-not (*M. alpestris*).
Forsythia is a genus of 6 species of shrubs. They have bright yellow flowers that appear in spring before the leaves open.
Foxglove (*Digitalis purpurea*) is a BIENNIAL herbaceous plant that produces a spike of purple flowers. The leaves contain the poisonous drug digitalin, used to treat heart disease.

Fuchsia is a genus of about 100 species of shrubs, found in Central and South America, and New Zealand. The flowers are red, purple or white and the coloured sepals surround a tube formed by the petals.

G Geranium is a genus of about 160 species of herbaceous plants. The flowers are white, pink or blue. Wild geraniums include herb robert (*G. robertianum*) and several kinds of cranesbill. Greenhouse geraniums have all been bred from the genus *Pelargonium*.

Geraniums

Gorse (*Ulex europaeus*) is a tough plant whose leaves are modified into spines. It belongs to the PEA family and has yellow flowers.

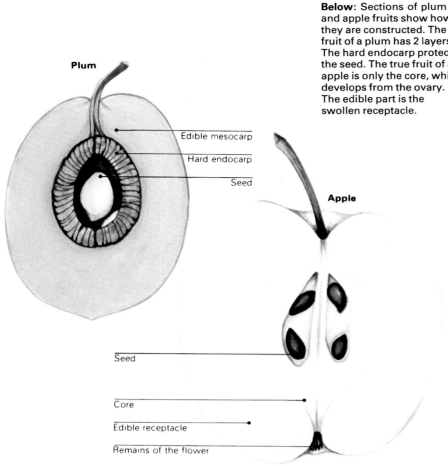

Plum

Edible mesocarp
Hard endocarp
Seed

Apple

Seed

Core

Edible receptacle

Remains of the flower

Below: Sections of plum and apple fruits show how they are constructed. The fruit of a plum has 2 layers. The hard endocarp protects the seed. The true fruit of an apple is only the core, which develops from the ovary. The edible part is the swollen receptacle.

Sometimes other structures, such as the petals, sepals, or receptacle are involved in the formation of a fruit. Such fruits are described by botanists as false fruits.

There are many different kinds of fruits. Each structure is related to the way in which the seeds inside are dispersed. As in the case of pollination, wind, water and animals all play a part in seed dispersal.

Wind dispersal is very common. The simplest form of wind-dispersed seed is found in some parasites and ORCHIDS. The seeds are small and light, and can be blown like dust. The poppy has a more elaborate mechanism. The seeds are contained in a capsule that sways in the wind. As it moves, the seeds are shaken out of tiny holes.

Some fruits have wings or parachutes. The wings of sycamore and ASH fruits spin like the blades of a helicopter rotor. The seeds are thus carried slowly to the ground. But as they fall the wind may carry them some distance from the tree. These wings have such excellent aerodynamic properties that they were studied by the pioneer aircraft builders in the 1800s. Parachuted fruits include those of the dandelion and old man's beard. The parachutes are made up of tiny hairs, and these fruits can be carried by the wind over long distances.

Dispersal by animals may occur in one of several ways. Sweet, succulent fruits, such as blackberries and plums, may be eaten by birds or mammals. They contain hard, indigestible seeds

Right: Fruits and seeds are dispersed in a number of ways. Dandelion and sycamore fruits are dispersed by the wind. Poppy seeds are shaken out of the holes in the top of the capsule. Laburnum seeds are ejected explosively as the pod twists open. Hazel nuts are hidden away by squirrels. Burdock fruits have hooks that catch on to the fur of animals. Strawberries are eaten by animals and the true fruits (achenes) on the outside of the swollen, red receptacle pass through the animal unharmed.

Fruit and seed dispersal

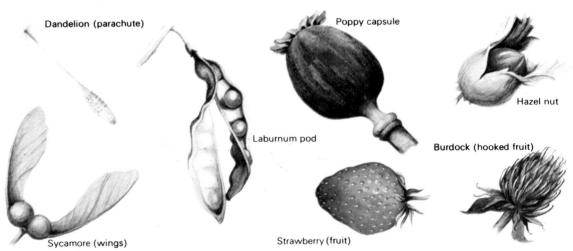

Dandelion (parachute)

Poppy capsule

Hazel nut

Laburnum pod

Burdock (hooked fruit)

Sycamore (wings)

Strawberry (fruit)

Grasses are a large family of monocotyledons. There are over 10,000 species including bamboos, sugar cane, and cereal crops, such as maize and wheat.
Groundsel are a few species of herbaceous plants belonging to the genus *Senecio*, which also includes the ragworts. They are all related to the daisy and have yellow flower heads.

H **Hazel** (*Corylus*) is a genus of 15 shrubs and trees found in northern temperate regions. The fil-

bert (*C. maxima*) and the European hazel (*C. avellana*) both produce edible NUTS.
Heartwood is the wood found in the centre of a tree

Honeysuckle

trunk. It has no living cells, and the xylem cells have become blocked, and so are not used for conducting water. It is more resistant to decay than SAPWOOD, and it gives the tree its strength. The pigments found in the heartwoods of ebony, rosewood and walnut make them very attractive for furniture.
Hemlock (*Conium maculatum*) is a very poisonous herbaceous plant related to COW PARSLEY, which it resembles. But it can be distinguished by its unpleasant smell and its smooth,

purple-spotted stem. A drug extracted from the leaves is now used in medicine.
Herbaceous plant is a plant without a woody stem.

Horse chestnut flower

Holly (*Ilex aquifolium*) is an EVERGREEN shrub with tough, shiny, prickly leaves. It is DIOECIOUS and the familiar red berries develop on the female tree when there is a male tree nearby.
Hollyhock (*Althaea rosea*) is a herbaceous garden plant, originally from Asia. It is a tall plant and bears many red, yellow, or white flowers.
Honeysuckle (*Lonicera*) is a climbing plant with heavily scented flowers that open in the evening. The wild honeysuckle, or woodbine (*L. periclymenum*), has white or

which pass undamaged through the animals. They may thus be dropped a long way from the parent plant. Hazel NUTS and oak acorns may be taken and stored by squirrels. In this case it is the absent-mindedness of the squirrels that ensures the dispersal of the seeds. They often forget where the nuts are hidden. Some dry fruits have hooks or spines on their surface. These catch on an animal's fur and are carried away.

Explosive mechanisms occur in several plants whose seeds are contained in pods or capsules. The two halves of a pea pod try to twist apart as its fibres dry out. Eventually, the strain bursts the pod, and the seeds are hurled out. Similar explosive mechanisms are found in *Laburnum* and *Geranium*. An unusual mechanism is found in the squirting cucumber. The seeds are forced out by fluid pressure, which builds up as the inner wall of the fruit breaks down.

Few fruits are dispersed by water, as the seeds are usually damaged. However, the fruits of the coconut have a thick, fibrous outer covering. This is waterproof, and the seeds may be carried

Above: Many animals eat the tasty fruits they can find. The seeds of a blackberry are sufficiently tough to pass through this dormouse without being harmed. This ensures widespread dispersal of seeds.

Left: Old man's beard gets its name from the mass of feathery fruits it produces in the autumn. They are dispersed by the wind.

over thousands of kilometres by ocean currents.

From seed to plant

After a seed has been dispersed it may begin to grow into a new plant. This is called germination. Most seeds need three things in order to germinate — water, light and warmth. Given these conditions some seeds germinate immediately. But other seeds may remain inactive for some time. This period in which the seed appears to do nothing is called dormancy, which literally means 'sleep'. Many seeds lie dormant during the winter months and germinate in the spring.

Germination begins when water enters the seed, causing it to swell. The embryo then begins to grow. First, it produces a tiny root, called the radicle. This grows down into the soil and helps to anchor the young plant. Next, the embryo produces a small shoot, or plumule. This grows upwards towards the light.

During this time the young plant is using the

---

pale pink flowers, which turn orange-brown after pollination. There are several cultivated species.
Hornbeam (*Carpinus*) is a genus of 21 species of deciduous tree found in northern temperate regions. The Eurasian hornbeam, or yoke elm (*C. betulus*), can grow up to 18 metres high.
Horse chestnut (*Aesculus hippocastanum*) is an ornamental tree thought to have come from Greece, but now found all over Europe and North America. Its white or pink flowers grow in large erect spikes.

Hyacinths are all members of the lily family that grow from bulbs. The wild hyacinth, or bluebell (*Endymion nonscripta*) is found all

Ivy in flower

over Europe. Cultivated hyacinths are all derived from the Lebanese species, *Hyacinthus orientalis*.
Hypogeal germination is a form of seed germination in which the cotyledons remain underground and play no part in photosynthesis. *See also* EPIGEAL GERMINATION.

I **Indehiscent fruit** is a fruit from which the seeds are not deliberately released. The seeds may germinate within the PERICARP, or they may be released by accidental breakage of the pericarp.

Iris is a genus of monocotyledons that includes over 200 species in the Northern Hemisphere. The outer segments of iris flowers are drooping; the inner segments are erect.
Ivy (*Hedera helix*) is a woody, evergreen climbing plant found throughout Europe. Other species are found in Asia and the Canary Islands. Ivy climbs by small roots that attach themselves to stone or bark by a cement-like substance.

J **Jacaranda** is a genus of 500 species of shrubs

Laburnum

and trees from Central and South America.
Judas tree (*Cercis siliquas-*

Below: A broad bean seed has its endosperm inside its 2 cotyledons. The cotyledons remain below ground during germination. This type of germination is called hypogeal. The endosperm supplies the developing plant with food until the first leaves appear. These then photosynthesize.

Below: A maize (sweetcorn) seed has a single cotyledon that remains below the ground all the time. The endosperm is not contained within it. The plumule is surrounded by a sheath called the coleoptile. The leaves develop inside this until they force their way through the top.

Below: A castor oil seed has its endosperm outside its cotyledons. But both are pushed above the ground during germination. This type of germination is called epigeal. The endosperm is absorbed by the cotyledons until they grow too large to be held within the seed case.

food reserves contained in the endosperm. In some seeds the endosperm is contained in the seed leaves, or cotyledons. In others the endosperm is separate from the cotyledons. In either case the seed may germinate in one of two ways.

During the germination of the sunflower seed, the cotyledons, which contain the endosperm, are pushed above the soil as the plumule grows. They quickly turn green and begin photosynthesis (*see page 77*). This also occurs when the castor oil seed germinates, although in this case the endosperm is not contained in the cotyledons. When the cotyledons emerge from the seed case, the endosperm is left behind.

In the other type of germination, the cotyledons remain underground. For example, the large cotyledons of the broad bean seed never emerge from the soil. They contain the endo-

Broad bean

Radicle

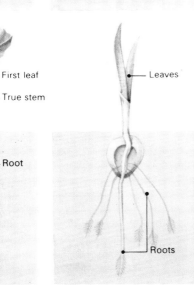

Plumule
Cotyledons — Radicle

First leaf
True stem
Root

Maize

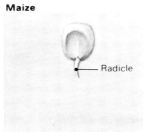

Radicle

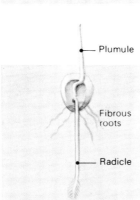

Plumule
Fibrous roots
Radicle

Leaves
Roots

Castor oil seed

Radicle

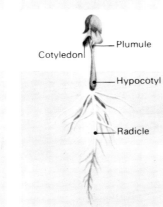

Cotyledon — Plumule
Hypocotyl
Radicle

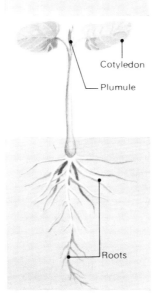

Cotyledon
Plumule
Roots

Below: A coconut may be carried for many thousands of kilometres by ocean currents before germinating on some tropical shore.

trum) is a beautiful deciduous tree found in southern Europe and Asia. It has pink flowers that open before the leaves in the spring. It is so called because Judas Iscariot is supposed to have hanged himself from one.

Laburnum is a genus of 3 small trees found in Europe and Asia. They belong to the PEA FAMILY, and produce seeds in pods.
Laurel family is a family of deciduous and evergreen shrubs including the bay laurel (*Laurus nobilis*), whose leaves are used in

cooking. The avocado pear and the sassafras are also in this family. The common laurel (*Prunus laurocerasus*) belongs to the ROSE FAMILY.

Common laurel

Legume is a pod or dry fruit formed from a single carpel, that opens by splitting down both sides. The term legume is often used to describe the type of plant that produces pods. For example, peas, beans, *Laburnum*, clover, and *Acacia* are all legumes.
Lily (*Lilium*) is a genus of about 80 species of monocotyledons found in northern temperate regions. They have large, showy flowers of all colours except blue.
Lily of the valley (*Convallaria majalis*) is a monocotyledon that grows in northern temperate regions. It has a

creeping underground rhizome, broad, pointed leaves, and white, bell-like flowers.
Lime (*Tilia*) is a genus of about 30 species of deciduous trees. Their soft, white wood is used for carving.
Loosestrife is a family of 50 species of trees, shrubs and HERBACEOUS plants. It includes the purple loosestrife (*Lythrum salicaria*), which is a herbaceous plant found in Europe. It has 3 different types of flower to ensure cross-pollination. The dye henna is obtained from the leaves of one of this family,

Lawsonia inermis, which is a shrub found in North Africa and south-west Asia.
Lords-and-ladies (*Arum maculatum*) is also known as

Lilies of the valley

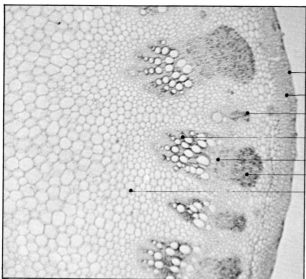

Left: A cross-section through a sunflower stem
Right: A cross-section through a maize stem.

Epidermis
Collenchyma
Parenchyma
Xylem
Phloem
Cambium
Schlerenchyma fibres
Parenchyma of pith
Parenchyma

Right: The annual rings can be seen on the cut ends of these larch trees. By counting annual rings on a tree trunk or stump you can tell the age of the tree.

sperm and therefore supply the growing plant with food. But they never take part in photosynthesis. The first green leaves are produced by the plumule. The single cotyledon of a maize seed also remains below the ground, even though the endosperm is outside the cotyledon.

Growing stems and roots

As a plant grows, its stem, branches and roots get longer. At the same time the whole plant becomes thicker and stronger. Growth occurs in special regions called meristems. The increase in the length of stems and branches occurs by growth of meristems in the buds. The tip of a root is also a meristem, and it is here that root growth occurs.

Inside a bud there is a region of dividing cells. As the cells divide, new cells are formed. The region of dividing cells thus moves forward, leaving behind some of the new cells. Further back from the tip of the meristem there is a region of differentiation — a region where cells become different from each other. They alter according to the function they must perform. For example, the cells on the outside of the meristem

the cuckoo-pint. This monocotyledon grows in northern temperate regions. It has an ingenious method of preventing self-pollination (see text).

Magnolia is a genus of evergreen and deciduous trees. Originally from Asia and North America, they are now widely planted in temperate regions. They have large, showy flowers, which may be pink, white, yellow or greenish.
Maple (*Acer*) is a genus of 60 species of deciduous trees found widely through-

out the Northern Hemisphere. They all produce 2-winged fruits. The leaves of the sugar maple (*A. saccharum*) of North America provide maple sugar. Other maples are popular garden trees. The sycamore (*A. pseudoplatanus*) is native to Europe. The Japanese maple (*A. palmatum*) has leaves that turn bright red in autumn. One of the tallest maples is the North American red maple (*A. rubrum*), which may grow to a height of over 35 metres.
Monoecious plant is a plant that has separate male and

female flowers on the same plant. *See also* DIOECIOUS.

Nettle (*Urtica*) is a genus of 30 species of

Ancient oak in winter

herbaceous plants. They have stinging hairs on their leaves. The stings of a few Indonesian species can be fatal. Stinging nettles are wind-pollinated and thus their flowers are greenish and indistinct. Deadnettles (*Lamium*) and hemp-nettles (*Galeopsis*) belong to a separate family and do not sting. They can be distinguished from stinging nettles by their purple or white flowers.
Nightshade family contains about 1,800 species of plants. Some of these are useful to man, such as the

potato, tomato, aubergine (egg-plant) and sweet pepper. However, others are very poisonous, such as deadly nightshade (*Atropa*

Passion flowers and fruit

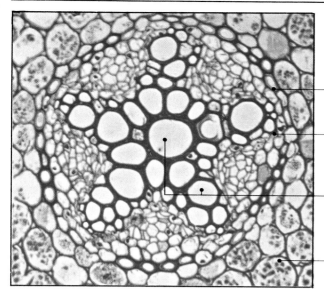

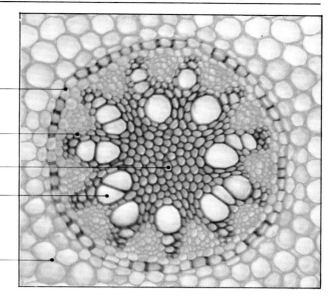

Left: A cross-section through a buttercup root. **Right:** A cross-section through a plantain lily root.

Endodermis

Phloem

Pith

Xylem

Cortex

become epidermal cells. Others become xylem, phloem or parenchyma cells (*see pages 72–74*).

The same process occurs in the tip of a root but here the region of dividing cells is protected by a covering of tough cells, called the root cap. This is needed because the root meristem would otherwise be damaged as it pushes down through the soil.

In fully-formed stems and roots the cells have become arranged in tissues. These are regions of a plant that have particular functions. For example, all the xylem vessels are arranged together into a water-conducting tissue, called simply the xylem.

The outside of a young stem is covered by the epidermis. Inside this is a region called the cortex, which may consist of parenchyma or collenchyma cells. The phloem and xylem are arranged together in bundles, called vascular bundles. In dicotyledons these vascular bundles form a ring inside the stem, and each bundle consists of phloem on the outside and xylem inside. In monocotyledons the vascular bundles are arranged at random. In the centre of the stem is an area called the pith, or medulla. This is usually made up of parenchyma cells.

The tissues of a root are arranged slightly differently. As in the stem there is an epidermis (called the piliferous layer) and a cortex but the xylem and phloem are placed in the centre of the root. The xylem is arranged like a star, with the phloem placed between the points.

Making wood and bark

If you look at the stump of a tree that has been cut down, you can usually see that it seems to be made up of rings. In fact, each ring represents the growth that took place in a single year. Thus, if you can count the rings you can tell the age of the tree when it was cut.

Each year's growth is due to the activity of yet another meristem (growth region) called the cambium. This is a layer of cells that forms a cylinder inside the stem of the woody plant. As the cells of the cambium divide, new xylem cells are formed inside the cylinder. These tough, thick-walled cells make up the wood of a tree. In the spring the cambium forms large xylem cells. As summer passes the new xylem cells become smaller until, in the autumn, the cambium stops producing cells altogether. Therefore, the rings that you see on a tree stump are made by the difference in appearance between the small autumn cells and the large spring cells.

The cambium also produces new phloem cells on the outside, but these are not as strong as xylem cells. As a result, each year's growth of phloem cells eventually becomes crushed, and so rings are not formed outside the cambium.

The epidermis that covers the outside of a young stem is not strong enough to protect a tree. Therefore, a layer of cells inside the epidermis begins to divide. This layer is known as the cork cambium. The new cells that are produced on the outside of this layer become thick-walled cork

---

belladona), woody night-shade (*Solanum dulcamara*), black nightshade (*Solanum nigrum*) and henbane (*Hyoscyamus niger*). The tobacco plant (*Nicotiana tabacum*) is also a member of this family.
Non-endospermic seed is a seed in which the endosperm (food store) is contained in the COTYLEDONS.
Nut is a one-seeded, INDEHISCENT FRUIT with a woody PERICARP. Examples include hazelnuts, sweet chestnuts, oak acorns and beechnuts. *Note:* some structures are incorrectly regarded as nuts. Coconuts and walnuts are

the stones of DRUPES; Brazil-nuts are seeds that come from a many-seeded, hard fruit (usually regarded as a

Plantains

BERRY); peanuts are seeds that come from a 2-seeded fruit.

O **Oak** (*Quercus*) is a genus of over 300 species of deciduous and evergreen trees. They are fine-looking trees, often with massive trunks, and may grow to over 30 metres high. Some of them are important timber trees, for example the North American white oak (*Q. alba*) and the English oak (*Q. robur*). Most of the large oak forests of England were cut down in the 1600s to build ships and houses. The

cork oak (*Q. suber*) is the main source of cork, which makes up its spongy bark.
Old man's beard, see CLEMATIS.
Orchids are a family of nearly 20,000 species of monocotyledons. They are found in all parts of the world except the coldest and driest regions, but most orchids are found in tropical areas. They have a distinctive flower structure in which one petal is extended to form a lip. Many orchids are epiphytes (*see page 115*). Many orchids, including the few British species,

Poplar

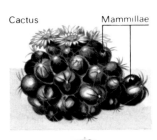

Cactus — Mammillae

Strawberry — Runner

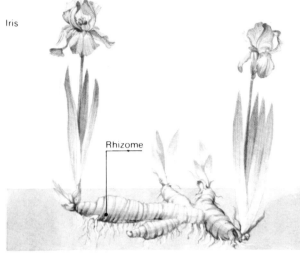

Iris — Rhizome

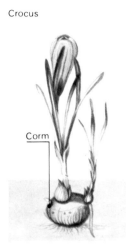

Daffodil — Bulb

Crocus — Corm

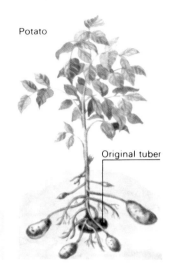

Potato — Original tuber

Above: Many plants can reproduce without using flowers. Some cacti produce growths, called mammillae, at their base. Strawberry plants produce long stems called runners. Irises grow from underground stems called rhizomes. Daffodils grow from storage organs called bulbs, which produce new bulbs each year. Crocuses grow from corms, which also produce new corms every year. Potatoes produce underground tubers. These are food storage organs and each one can grow into a new potato plant.

cells, and form the bark of the tree.

Other ways of multiplying

The process of producing a seed involves sex cells. This is the most important way in which flowering plants reproduce, as it results in strong, healthy plants. Also, new varieties may be produced, and eventually new species may evolve.

However, there are various ways in which new plants are formed without sex cells being involved. Various parts of plants are specially modified for this purpose. Usually, it is the stem that is modified.

Rhizomes, such as those of the IRIS, grow underground. They are not roots, they are modified stems. They grow horizontally under the soil. At intervals they grow shoots, which develop into the visible parts of the plants. The rhizomes of the potato plant are much thinner but at certain points they swell up into food storage organs called tubers. Shoots spring up from little buds, called 'eyes', on the surface of the tuber and roots grow from other 'eyes'. The

shoots develop into a bushy plant which sends out side stalks close to the surface where new tubers develop.

Runners, such as those of the strawberry plant, are also modified stems. But they grow above the soil. Where they touch the ground new roots are formed. When the new plant is established, the runner withers away. Gardeners find this a useful way of growing new strawberry plants.

Bulbs, such as DAFFODILS, are also modified stems. In fact the true stem is only the solid heart of the bulb. The rest of the bulb is made up of fleshy leaves. New bulbs grow from one or two of the buds that form where the leaves join the stem. A corm, such as that of a CROCUS, is similar to a bulb but it consists mainly of a swollen stem. The leaves are reduced to scales. At the end of the year the old corm dies and a new corm grows above it. At the same time the old corm may produce one or two smaller new corms. Corms and the leaves of bulbs are also food storage organs. The food is used to help the new growth of the plant in spring.

The good-luck plant (*Bryophyllum*) has an

are rare and therefore no orchids should be picked.

P Palms are a family of about 3,500 species of monocotyledons found in the tropics and sub-tropics. Many species are important to man, including the coconut palm (*Cocos nucifera*), the date palm (*Phoenix dactylifera*), the West African oil palm (*Elaeis guineensis*), and the Malaysian sugar palm (*Arenga saccharifera*).

Pansy, see VIOLET.

Passion flower (*Passiflora*) is a genus of more than 300 species of climbing plants,

originally from the tropical regions of America. These plants got their name because their flowers were thought to represent the passion of Christ. Inside the brightly coloured sepals and petals there is a ring of thread-like filaments, and the 3 stigmas are arranged in the shape of a cross.

Pea family is a family of over 7,000 species of plants. They have butterfly-like flowers and produce their seeds in pods. Members of this family include ACACIA, *Mimosa*, LABURNUM, GORSE, peas and beans.

Perennials are plants that continue to grow from year to year. HERBACEOUS perennials die down during the winter and produce new

Rowan tree in flower

growth from their underground parts in the spring. Woody perennials have permanent woody parts above ground all year round. *See also* ANNUALS, BIENNIALS.

Perianth is the CALYX (sepals) and COROLLA (petals) of a flower. This term is generally used to describe the outer parts of flowers in which the petals and sepals are difficult to tell apart; e.g. the flowers of tulips, irises and lilies.

Pericarp is the outer layers of a fruit, all derived from the wall of the original ovary.

Pink family is a family of

over 1,000 species of herbaceous plants. They include carnations, pinks and sweet williams, which all belong to the genus *Dianthus*. Also included are the campions (*Silene*), stitchworts (*Stellaria*), and chickweeds (*Cerastium*). Their flowers are pink or white.

Plane (*Platanus*) is a genus of 6 species of deciduous trees. They are often planted in towns and cities because they are resistant to smoke and fumes.

Plantain (*Plantago*) is a genus of herbaceous plants

Right: Many climbing plants are assisted by tendrils. Young tendrils grow straight until they touch a support. If they do not find a support, they eventually coil up and are not used. After a tendril has coiled round a support, the part of the tendril nearest the plant coils in the opposite direction to the part nearest the support. This creates tension and draws the plant and the support closer together.

Below: Plants always grow towards the light. Even if this pot were turned round the seedlings would bend back towards the light. This movement is called phototropism.

unusual way of producing new plants. Along the edges of its leaves it grows tiny plants called bulbils. These fall off, and each one can grow into a new plant.

All these ways in which plants reproduce are called vegetative propagation, and they occur naturally. But man has invented some artificial methods of producing new plants from old ones. The simplest of these is called taking a cutting. A small shoot can be cut from a young stem and planted in soil. After some time it may grow new roots. Chrysanthemums, geraniums and lupins are just three of the many plants from which successful cuttings can be taken. Woody shoots, such as those from rambling roses and gooseberry bushes, can also be treated in the same way. Some plants, such as Begonias, will grow from leaf cuttings.

However, a number of plants cannot be grown from cuttings. In such cases, gardeners use another method called grafting. For example, new fruit trees can be grown from grafted twigs. First, a young, healthy sapling is selected. This is often a wild variety. The main stem is cut off just

---

found all over the world. They have small, colourless flowers on long spikes. Because plantains will grow in all soils and locations, they are often troublesome weeds on paths and lawns.

Pod, see LEGUME.

Poplar (*Populus*) is a genus of about 35 species of deciduous trees related to the WILLOWS. They are found all over the Northern Hemisphere. Their male and female catkins are borne on separate trees. This genus also includes the aspens and American cottonwoods.

Poppies are a family of about 450 species of herbaceous plants found in subtropical and northern temperate regions. British species include the field poppy, or corn poppy (*Papaver rhoeas*), and the yellow horned poppy (*Glaucium flavum*) which grows by the sea.

Primrose (*Primula vulgaris*) is a herbaceous plant that flowers in spring. It grows in woods, hedgerows and grassy places. Pink, purple or white varieties, as well as the pale yellow type, occur.

Privet (*Ligustrum vulgare*) is an almost evergreen shrub found all over Europe. Its white clusters of flowers have an unpleasant smell, and its black berries are poisonous to eat.

Snowdrop

R Rhododendron is a genus of over 500 shrubs that come from East Asia. They are all evergreen, except for one group known as azaleas. They have brightly coloured flowers, which are often scented. They are popular shrubs for parks and gardens, but they will not grow in lime-rich soils.

Rose family is a family of about 2,000 species of trees, shrubs and herbaceous plants. Members of this family include plums (*Prunus*), apples (*Malus*), hawthorn (*Crataegus*), blackberry (*Rubus*), and strawberry (*Potentilla*), as well as the true roses (*Rosa*). The 150 species of wild rose are mostly found in the Northern Hemisphere.

Rowan tree (*Sorbus aucuparia*) is a deciduous tree, also called the mountain ash, found in the Northern Hemisphere. It has bright orange-red berries in the autumn.

S Sapwood is the outer ring of wood surrounding the HEARTWOOD of a tree. It contains some living cells, as well as xylem cells, and conducts water up the tree.

above the roots. The rootstock that remains will form the roots of the new tree. A twig from a tree of the desired variety is then inserted into the bark at the top of the rootstock. This twig grows and eventually forms the flowers and fruit that the gardener requires.

ROSES are propagated by a special method of grafting, called budding. In this case, instead of a twig, a bud is inserted into the top of the rootstock.

Plant movements

Though plants do not move from place to place, certain movements do occur, and they are all concerned with the need for water, food, light, protection or support.

If you grow some seeds indoors near a window, you will find that all the seedlings lean towards the light. Even if you turn the container round, within a few hours the seeds will bend back towards the window. This movement is called phototropism. Because plants must have light in order to make food, they have developed a system that enables them to detect and grow towards a source of light.

A stem always grows upwards, and a root always grows downwards. You can prove this by taking a bean seedling that has already produced a shoot and root. Turn it upside down so that the

Above and below: *Mimosa* is a plant that has sleep movements. In the normal position *(above)* the leaflets are erect. In the sleep position *(below)* the leaflets are folded down. This normally occurs in the evening, but a sudden tap can cause it in the day.

stem points downwards. Within a few days the root will have curled downwards, and the shoot will have turned so that it once again grows upwards. You can do this in the dark, so that phototropism is not the cause. In fact it is gravity that the plant responds to, and the movement is called geotropism. This is an important movement because it ensures that the root of a seedling grows down into the soil where there are minerals and water. At the same time the shoot grows up through the soil, eventually reaching the light.

There are a number of other plant movements. Some flowers, such as those of the DAISY, open during the day and close at night. Crocus flowers open only when the temperature rises above a certain point. Some climbing plants have tendrils that grow out in spirals. When a tendril finds a support, it quickly coils round it. Some bean plants of the genus Vicia have a 'sleep movement' – a raising and lowering of the leaves from the horizontal during the day to the vertical at night.

One of the most spectacular plant movements is shown by *Mimosa*. During the day the leaves are erect, but at night the plant appears to wilt. This sleep movement can also be made to happen by giving the plant a sharp knock. The leaves collapse suddenly, due to a very rapid change in turgor pressure (*see page 74*).

Schizocarp is a fruit that breaks into several pieces, which are generally one-seeded. Examples include the fruits of *Geranium*, dead-nettles, and hemlock.

Shrub is a woody plant that has several stems arising from the soil. *See also* TREE.

Snapdragon, see ANTI-RRHINUM.

Snowdrop (*Galanthus*) is a genus of 13 species of monocotyledons found in Europe and the Middle East. They grow from bulbs and have bell-shaped flowers.

Sunflower (*Helianthus annuus*) is an annual her-baceous plant related to the daisy. The flower heads have many brown disc-florets surrounded by a ring of large, yellow ray-florets.

T **Thistles** are a group of herbaceous plants re-lated to the daisy. Their flower heads are made up of purple, tubular disc-florets. British species include the common field thistle (*Cirsium arvense*), the marsh thistle (*C. palustre*), the slen-der thistle (*Carduus tenuif-lorus*), and the Scotch thistle (*Onopordum acanthium*).

Tree is a woody plant that

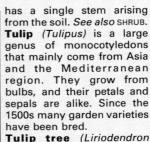
Scotch thistle

has a single stem arising from the soil. *See also* SHRUB.

Tulip (*Tulipus*) is a large genus of monocotyledons that mainly come from Asia and the Mediterranean region. They grow from bulbs, and their petals and sepals are alike. Since the 1500s many garden varieties have been bred.

Tulip tree (*Liriodendron tulipifera*) is a deciduous tree from North America. It is related to the magnolias and has large, greenish flowers.

V **Vines** are a family of climbing plants, mostly

from tropical and sub-tropical regions. They in-clude the grape (*Vitis vin-ifera*).

Violet (*Viola*) is a genus of herbaceous plants that in-cludes the pansies. There are about 400 species, mostly small herbaceous plants. They grow all over the world.

W **Willow** (*Salix*) is a genus of about 170 de-ciduous trees found all over the world. They are graceful trees, and the Chinese weep-ing willow (*S. babylonica*) is a popular garden variety.

Plants have adapted to cope with all kinds of climates: hot and cold, dry and swampy, salt water and fresh water. Each plant is suited to its own environment and this accounts for the enormous variety of plant life on the Earth.

Adaptation to Environment

Left: Mangroves have stilt roots to support them in the swampy ground in which they grow. This genus (*Rhizophora*) has breathing pores in the bends of the stilts to provide oxygen to the roots below.
Below: Air spaces in the leaves and stem of a water lily help the leaves to float. They also help to supply oxygen to the plant.

Air spaces

Plants need water and food to survive no matter where they live. In most temperate climates these are present in suitable amounts. Therefore, it is relatively easy for plants to take in what they need. The adaptations of plants to such environments are those that we accept as the normal plant structures – a standard root, stem and leaf system.

However, there are climates that are less hospitable. For example, water may be scarce, or there may be too much of it. The plants that live in normal environments cannot survive under such conditions. However, plants, and particularly flowering plants, are very adaptable. Many of them have evolved ways of overcoming hardships. Thus even the inhospitable environments of the world have their own particular plant populations.

Living in water

Since water is essential for plants, it may seem surprising that a plant can have too much water. After all, many algae manage to live permanently submerged in water. Ferns, gymnosperms and flowering plants are basically land plants which evolved from plants that established themselves on land millions of years ago. Only more recently have some of them returned to a watery environment – like the whales in the animal kingdom.

A land plant needs to have a strong stem for support but an aquatic (water) plant is held up by the water around it. Thus, one of the characteristics of such plants is the reduction of the stem tissues. They have few fibres and little or no xylem. Instead, they have air spaces that help to keep them floating in the water. These air

---

Reference

A **Alpine zones** occur on mountains below the permanent snow fields and above the timber line. This type of climate is cold and harsh and supports only hardy, low-growing plants.
Amazonian water lily (*Victoria amazonica*) is the largest of all the water lilies; its leaves may be up to 150 cm across, and its flowers are about 30 cm across.
American century plant (*Agave americana*) is a

North American desert plant. It has long, tapering, fleshy leaves covered in a tough cuticle. It gets its name from the fact that it may take over 50 years to flower.
American resurrection plant (*Selaginella lepidophylla*) (*see page 94*) that can withstand being completely dried out. Although it appears to be rolled up and dead, when moistened it unrolls and revives. It is often sold as a curiosity.
Azolla is a fern that grows in water. The undersides of

its fronds are covered in special hairs that repel water. As a result the fronds float, and dense mats of this fern can cover ponds. The

B **Banyan** (*Ficus*) is a genus of Asian and Afri-

fronds turn red in the autumn.

can plants related to the fig. Banyans begin life as EPIPHYTES on palm trees. They put down aerial roots, which eventually reach the soil. The roots are able to fuse together round the host tree. Finally, the host is strangled, and the banyan remains, a tree with a 'trunk' made of aerial roots. Banyans also spread from tree to tree along the branches, putting down aerial roots all the time.
Bindweed *see page 101.*
Bird's nest orchid (*Neottia nidus-avis*) is a plant whose leaves contain little or no

Amazonian water lily

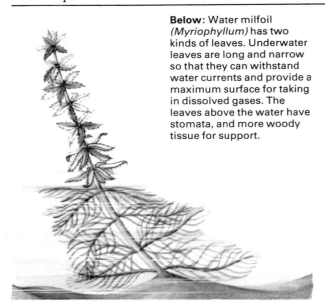

Below: Water milfoil *(Myriophyllum)* has two kinds of leaves. Underwater leaves are long and narrow so that they can withstand water currents and provide a maximum surface for taking in dissolved gases. The leaves above the water have stomata, and more woody tissue for support.

spaces also perform another function. They help oxygen to reach the underwater parts of the plants.

Submerged plants in rivers, like the Canadian pondweed, have to survive the battering they receive from the water currents. To do this they have flexible stems and long, thin leaves. Water currents pass smoothly over these plants, and the leaves do not tear.

Aquatic plants with floating leaves have the problem of avoiding damage by wind and waves. The best-known of these plants are the water lilies. They have circular leaves, and scientists think that this is probably the best design for avoiding the possibility of being torn or swamped by waves. The largest example of this type of plant is the giant AMAZONIAN WATER LILY, whose leaves may be one and a half metres across. They are reinforced by stiff ridges underneath and are strong enough to take the weight of a child.

There are many other plants with floating leaves and they are generally found in sheltered ponds, where damage is less likely to occur. Some of them, such as the duckweeds, are free-floating. A single duckweed plant consists merely of a simple leaf-like body and one or more hanging roots. WOLFFIA, the smallest of all the flowering plants (less than half a millimetre across), consists of a tiny round body with no roots, stem or leaves.

Plants with parts above the water may be little

different to land plants, particularly if they live in only shallow water. The common reed is an ordinary monocotyledon. However, plants that live in deeper water may have more complicated structures. MANGROVES live in coastal swamps. They have tall, arching stilt roots that keep the main parts of the plant clear of the high tide mark, whereas their roots are buried in mud that contains little oxygen. Therefore, they have specialized roots called pneumatophores that project above the mud and water to get oxygen from the air.

Living in salty conditions

The problem of a plant that lives near the sea or an inland salt lake is how to get the water it needs. It is a problem because the water outside the plant contains a high concentration of salt (and is therefore a strong solution). Osmosis, the process by which plants take in water, works by moving water from a weak solution to a strong solution (*see page 75*).

The simplest way to overcome this problem is to ensure that the cell sap is stronger than even the salty water outside. Thus osmosis can still take place. Most plants that live in these conditions have much stronger cell solutions than plants living under normal conditions.

However, there are times when even this does not work. For example, after a storm there may be much more salt than usual in the soil. To

Above: The rhizomes of marram grass help to bind sand dunes together. Marram grass can survive the dry conditions on the dunes because of the structure of its leaves.

---

chlorophyll (green pigment). Hence it cannot make its own food. Many people regard this plant as a saprophyte (feeding on dead organic matter). In fact it has a mass of short thick roots that are entangled with a fungus. Such a relationship is usually either a symbiosis, or the fungus is a parasite but in this case the orchid relies on the fungus to break down the organic matter on which it lives. It cannot do this by itself, and it can therefore be said to parasitize the fungus.
Bladderworts *(Utricularia)*

are carnivorous plants found in water and in damp places, particularly in tropical regions. They have tiny bladders that trap small animals. Each bladder is closed by a small valve. Inside the bladder there is a slight vacuum. When an animal disturbs the hairs around the opening, the valve flies open and the animal is sucked into the bladder. The valve closes and the animal dies in the bladder. Gradually, the animal is digested, and this process helps to create a new vacuum inside the bladder. Thus the trap is re-set.

Bird's nest orchid

Bramble *see page 121.*
Bromeliads are a family of monocotyledonous South American plants, many of which are EPIPHYTES. Some of them catch water in cups formed by the wide bases of their leaves.
Broomrapes *(Orobanche)* are a genus of parasitic plants found in Europe, Asia and Africa. Their leaves contain no chlorophyll and their tubers are attached to the roots of host plants, such as clovers, peas and daisies.
Butterworts *(Pinguicula)* are a genus of carnivorous plants. They have small

rosettes of leaves covered with a butter-coloured, sticky substance that traps insects.

C **Cactus** *(plural:* cacti) is a member of the family Cactaceae. They are all succulent plants with modified stems for withstanding drought conditions. Their leaves are modified into spines, hairs or bristles.
Californian poppy *(Eschscholtzia californica)* is an EPHEMERAL desert plant found in North America.
Canadian pond weed *(Elodea canadensis)* is a

Above: After a rainstorm this area of desert in Central Australia is blooming with ephemeral (short-lived) flowers.

Left: The filamentous alga *Spirogyra* forms a green scum on the surface of a sheltered pond. The lesser duckweed *(Lemna minor)* floats on the surface, together with the reddish leaves of the water fern *Azolla*.

Right: The glasswort *(Salicornia)*, with its fleshy leaves, can live in the salty waters of an estuary.

overcome this many seaside plants, such as the glassworts, have fleshy leaves that store water. In this way they can survive until the concentration of salt around them becomes low enough again.

Living in dry conditions

The deserts provide some of the harshest climates to be found in the world. Some places, such as the middle of the Sahara desert, are so dry that nothing can live there. However, given even a little water some plants can survive, using their special adaptations.

The simplest way to combat drought is to avoid it altogether. Parts of some deserts are subject to long periods of drought, followed by sudden short periods of torrential rain. During the drought, few plants appear to be present. But immediately after the rain a large number of plants emerge from seeds that have been lying

---

North American water plant, now abundant in Britain. It grows completely submerged and its elongated leaves are borne on long, flexible stems.

Candle plant *(Kleinia articulata)* is a stem succulent related to the daisy. In times of drought it loses its few leaves and survives on the water stored in its swollen stem.

Creosote bush *(Covillea glutinosa)* is a North American desert plant. It looks like an ordinary plant, but it is able to withstand drought because its cells can still carry out their work when they have lost water.

D **Deserts** are areas of land that receive little or

Cacti in France

no rain. Some are permanently hot, such as the Sahara desert in North Africa but the northern deserts of Asia are extremely cold at night and during the winter.

Dodder *(Cuscuta)* is a genus of parasitic plants found in many parts of the world. They are parasites of clovers, hops, nettles, gorse and heather, and may cause damage to crops.

Duckweed *(Lemna)* is a genus of small free-floating water plants found on ponds, ditches and lakes all over the world.

E **Eelgrass** *(Zostera)* is one of the few water plants (except for seaweeds) that lives in the sea. It is submerged and even produces

Stony desert

underwater flowers.

Ephemeral plants are those that only live for a very short period of time – just long enough to flower and produce seeds.

Epiphytes are plants that grow on other plants. They are not parasites, they only use their hosts for support.

G **Glasswort** *(Salicornia)* is a succulent plant that lives near the sea in salty conditions.

H **Halophytes** are plants that grow in salty soil.

Honeysuckle *see page 105*.

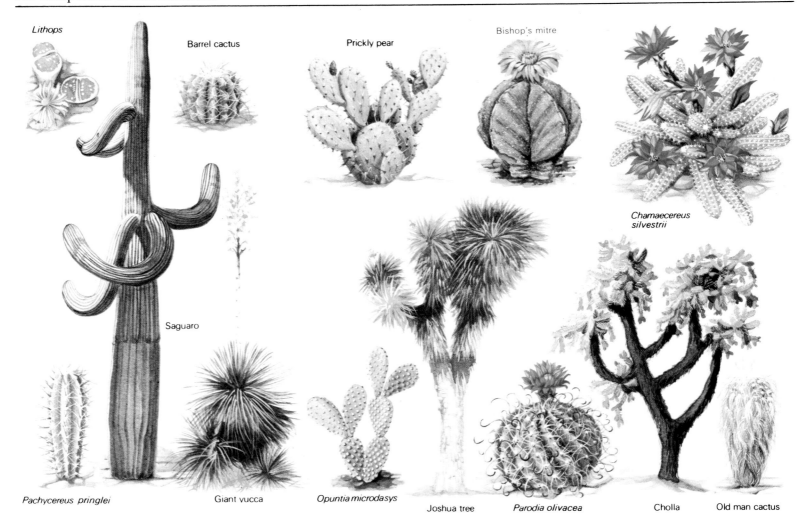

Lithops

Barrel cactus

Prickly pear

Bishop's mitre

Chamaecereus silvestrii

Saguaro

Pachycereus pringlei

Giant yucca

Opuntia microdasys

Joshua tree

Parodia olivacea

Cholla

Old man cactus

dormant. These are called ephemeral plants. They grow, flower and produce seeds in a very short time, before the heat kills them. The new seeds remain in the soil, waiting for the next rainstorm.

Some plants are able to endure the loss of water in dry conditions. The CREOSOTE BUSH, which is found in the deserts of North America, looks like a normal plant but its cells can withstand being dried out. The AMERICAN RESURRECTION PLANT is another curious example. In dry conditions it rolls up and appears dead but when it becomes moist again it unrolls. Many mosses are also able to withstand drying out. *Tortula muralis*, for example, a common wall moss in Britain, revives rapidly after a shower of rain.

However, most plants cannot endure water loss, and so those that live in dry conditions have to have some means of preventing it. The first and most obvious step to take is to reduce the rate of transpiration. MARRAM GRASS, which grows on sand dunes and is common in Britain, has an excellent method of doing this. Its leaves are covered with a tough epidermis. They are also curled, and the stomata are situated in pits. A moist atmosphere is thus created above the stomata, and transpiration is slowed down.

Plants in hotter climates require more drastic measures, and the best-known desert plants are those that store water – the cacti and other succulents. Some of these succulents store water in their leaves, such as the pebble-plants ('living stones') and the AMERICAN CENTURY PLANT.

Above: These cacti flourish in the deserts of the American continent. They are all adapted to living in dry, arid conditions and have fleshy stems or leaves for water storage. Some have brightly coloured flowers which attract insect pollinators.

Hydrophytes are plants that grow in water.

I **Ivy** see page 106.

L **Lianas** are climbing plants found in tropical forests. They have woody stems. Some climb by twining round trees. Others use tendrils, or thorns.

M **Mangroves** include several shrubs that belong to 3 different families. However, they all have stilt roots and pneumatophores (adapta-

tions for life in tidal mud-flats).
Marram grass (Ammophila arenaria) grows on raised

Marram grass

sand dunes and cannot tolerate immersion in the sea (unlike SEA COUCHGRASS). The binding action of its roots

and rhizomes cause larger dunes to build up. Eventually the dunes become more stable and contain more humus (from dead marram grass). At this stage marram grass is succeeded by other plants. Sand dunes contain very little water, and marram grass is adapted to these dry conditions. Its curled leaves and sunken stomata help to slow the transpiration rate.
Mistletoe is a partial parasite of trees. The common mistletoe is usually found on apple trees. In some parts of the world mistletoe can be a serious tree pest.

O **Old man of the desert** (Cephalocereus senilis) is a Mexican cactus that has silvery, hair-like outgrowths instead of spines.

P **Papyrus** (Cyperus papyrus) is a reed-like water plant found in Africa. It is normally rooted beneath the water, but sometimes forms vast floating rafts. In ancient Egypt papyrus was used for building and papermaking.
Parasite is a plant or animal that lives at the expense of another plant or animal, without giving anything in return.

Others, such as cacti and the candle plant, use their stems as storage organs. The main purpose of the shape of these plants is to reduce the surface area exposed to the drying air. The outer layers of succulents are tough, and there are few stomata. As a result the rate of transpiration is very slow. In this way some cacti can survive for several years without water.

The bristles and spines of cacti are a further adaptation. They are actually modified leaves, and they serve to discourage animals in search of a juicy meal. The bristles of the prickly pear cactus break off easily and act like itching powder.

Living with other plants

The majority of plants live more or less independently, although members of a plant community may get some benefit from each other, such as protection. Most of the plants in such a community make their own food and support themselves. However, some plants cheat and allow others to do some or all of the work for them. Such plants are specially adapted for obtaining their food or support, or both of these from other plants.

Sometimes two plants may live in a close association that is beneficial to both plants. Each one provides something that the other needs. This kind of relationship is called symbiosis. Many fungi form symbiotic relationships (*see page 89*). For example in a lichen, the fungus gains food from the alga, and the alga gets protection

Below: A cactus survives in the desert because it stores water in its stem. The spines, which are modified leaves, protect it from animals.

Above: *Lithops bella,* a pebble plant, stores water in its leaves. It is overlooked by animals because of its resemblance to pebbles.

Below: The creosote bush can survive in places where no other plants exist. Its cells are able to withstand being dried out.

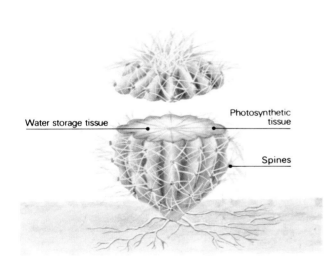

Water storage tissue

Photosynthetic tissue

Spines

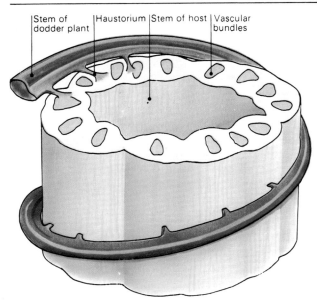

Stem of dodder plant | Haustorium | Stem of host | Vascular bundles

Above: Dodder is a total parasite, relying on its host for all its food and water. It coils round and draws food from the host's xylem and phloem.

Below: Mistletoe is only a partial parasite. It takes water and some food from its host, but its leaves are green and are able to photosynthesize.

Below: A Chinese banyan (*Ficus retusa*) has a large 'trunk' made of aerial roots. As the branches grow outwards, more aerial roots are put down.

from bright light and water loss. The symbiosis of a fungus with the roots of a flowering plant is called a mycorrhiza. The fungus takes sugar from the roots, and the flowering plant benefits because the fungus is more efficient in absorbing water and minerals from the soil.

There are also plants that cheat even more and rely entirely on others for their existence. Such plants are called parasites. There are many examples of parasites among the fungi, but there are also some flowering plant parasites.

MISTLETOE is a parasite of trees. It attaches itself to the branches and takes water from the xylem tissue of the host. It is however only a partial parasite, as it has green leaves and is able to make its own food.

Several flowering plants are total parasites. For example, DODDER is a parasite of clover, nettles, hops, gorse, and heather. In America one species twines itself round the stems of lucerne and alfalfa and makes these crops difficult to harvest. When a dodder seed germinates, it first puts down a small root. Then the shoot grows rapidly upwards. If it finds a host, the shoot curls

S **Saguaro cactus** (*Carnegiea gigantica*) is a very large cactus that forms forests in the deserts of Arizona, USA.

Salvinia is a fern that lives in water. Its leaves form star-shaped patterns on the surface of the water. It is kept afloat by air trapped in-between tiny hairs on its undersurface. *Salvinia* has no roots, but feathery leaves hang down below the surface of the water. It originated in Central and South America, but is now established in Central Africa and Sri Lanka, where it is a serious problem. It grows rapidly, forming dense mats that choke the waterways.

Sarracenia is a genus of 10 species of carnivorous plants found in North America. They are similar to pitcher plants, but their tall, trumpet-shaped pitchers grow directly out of the ground. Insects fall into the liquid inside the pitcher because it is lined with a special epidermis on which they cannot walk. This is the same type of epidermis as the one found lining the flower head of an arum plant (*see page 104*).

Scarlet pimpernel

Scarlet pimpernel (*Anagallis arvensis*) is a herbaceous plant related to the primrose. It is a common weed found in gardens, fields and roadsides. In such conditions it has normal leaves but when the plant grows in salty conditions, the leaves become succulent. Thus the scarlet pimpernel is an example of a plant that can adapt to the conditions around it.

Sea couchgrass (*Agropyron juncaeum*) is a grass that can tolerate being immersed in sea water. It therefore grows near the edge of the sea. Its deep, extensive system of rhizomes helps to bind the sand together. As a result, small sand dunes form. This is the beginning of natural land reclamation. Sea couchgrass is succeeded by other grasses, particularly MARRAM GRASS.

Sedum is a genus of plants with succulent leaves. It can therefore survive dry conditions. This genus includes the stonecrops and several houseplants grown for their attractive leaves.

Subtropical regions are those where it is warm in both winter and summer.

round the host's stem and puts out suckers called haustoria. These penetrate the host's tissues, and the dodder plant can then get water and food from the host's xylem and phloem. The root of the dodder plant dies, leaving it completely dependent on its host.

Other flowering plant parasites include BROOMRAPES, TOOTHWORTS and RAFFLESIA, which parasitize the roots of their hosts.

Some plants cheat in the way that they get light for photosynthesis. In a wood or jungle, most plants put a lot of energy into growing thick stems in order to reach the light. Climbing plants, however, save this energy by using other plants for support. They are not parasites, because they have underground roots and green leaves, but they cannot grow properly without the support of trees or walls.

Various methods are used for climbing. The

Above: A bromeliad is an epiphyte, living high in the branches of trees. It traps water in a central 'tank' and in hollows at the base of its leaves.

Below: The European sundew *(Drosera rotundifolia)* is a carnivorous plant that traps insects in the sticky fluid exuded from its tentacles.

bindweeds curl anticlockwise up the stem of a supporting plant; honeysuckles curl clockwise. Grasping climbers, such as the sweet pea, have tendrils that they use for holding on to their support. The virginia creeper has tendrils with suckers, which stick to the surface over which the plant is climbing. Penetrating climbers, such as ivy, have roots that can grow into the bark of a tree or the surface of a wall. Brambles do not make a positive attempt to climb. But they grow long shoots that fall over when they get too long. The sharp prickles catch on to sturdier plants, and in this way the plants ramble over the surrounding vegetation.

Another group of plants have an even simpler method of getting nearer the light. They grow high up in the branches of trees. Again, they are not parasites because they take no food from the trees. They get their food from rotting leaves and other material trapped in the branches. These plants are called epiphytes. In temperate climates the only epiphytes are mosses and ferns, but in tropical areas, epiphytes include many orchids and BROMELIADS.

The main problem for epiphytes is how to get water. They solve this by putting down aerial roots, which absorb water from the humid air of the tropical forest.

Carnivorous plants

Plants that live in marshy ground, such as peat

They extend outside the tropical regions to about 36°S and 36°N.
Succulent plants are those that have swollen, water-storing leaves or stems. Leaf succulents include PEBBLE PLANTS, SEDUM, and the AMERICAN CENTURY PLANT. Stem succulents include cacti and the CANDLE PLANT.
Sundews *(Drosera)* are 3 species of carnivorous plants with tentacles on their leaves. They grow in places where the soil is poor, such as peat bogs. When an insect lands on a sundew leaf, it is caught by the sticky

droplets on the ends of the tentacles. These bend over to hold the insect more firmly against the leaf. They then secrete a liquid that digests the insect. *D. rotundifolia* has spoon-shaped leaves that spread out near the ground. *D. anglica* has longer, more tapering leaves on upright stalks. *D. intermedia* has very long leaves.

T Temperate regions are those that have a warm summer and a cold winter. Most temperate land areas are in the Northern Hemisphere and extend from the

Toothwort

subtropics to the TUNDRA.
Toothworts *(Lathraea)* are a genus of parasitic plants found in Europe and Asia. They are related to the BROOMRAPES and are parasitic on the roots of trees. The British species *(L. squamaria)* is usually found on the roots of hazel or elm trees. It has no surface from which transpiration can take place. The water that it takes up from its host is exuded through special water glands. *L. clandestina* has no stem above the ground, and some of its bright purple flowers may remain below

the soil.
Tortula muralis is a very common moss on walls. It forms neat cushions, and its leaves are tipped with long, silvery hairs. Its spore capsules are pointed and upright. Its cells are able to withstand water loss. In dry conditions the moss may appear to be dead, but it quickly revives after rain.
Tropical regions extend between 2 lines of latitude – the Tropic of Cancer (23° 27'N) and the Tropic of Capricorn (23° 27'S). In this region the Sun is overhead twice a year, and the weath-

Above and below: Pitcher plants, such as *Nepenthes madagascariensis,*grow in tropical rain forests, either as epiphytes or in the soil. Insects are attracted by its scent. They cannot hold on to the shiny surface and therefore fall into the fluid at the bottom of the pitcher.

Right: The defences of some nettles are formidable. *Urtica ferox,* from New Zealand, is a woody shrub with large stinging hairs.

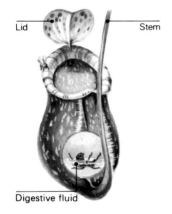

bogs, have difficulty in getting all the food they need. Some plants have therefore adapted to this problem by catching and digesting insects. They are thus the killers of the plant world.

There are a variety of techniques that plants use to catch their prey. The most spectacular carnivorous plant is the VENUS FLY TRAP. Its two-lobed leaves are fringed with spikes. When an insect lands on a leaf, the lobes close, trapping the insect inside. A pitcher plant, which may live as an epiphyte or in swampy ground, has a vase-like construction. Insects, attracted by the nectar on the lid, fall into a digestive fluid at the base, from which they cannot escape. The tentacles of a SUNDEW plant are tipped with drops of sticky liquid. These trap any small insects that land on the plant. The leaves of butterworts are completely covered in a sticky substance, which glues

the insects down while they are being digested. BLADDERWORTS are plants that trap small water animals in their bladders. The opening of a bladder is closed by a small valve. When an animal touches the hairs round the opening, the valve opens and the animal is sucked in.

Plant defences
Many plants have developed ways of making themselves less attractive to plant-eating animals. Some produce poisons that cause an animal to be sick, and this discourages the animal from eating that plant again. A nettle produces poison in the hairs on its epidermis. A hair injects the poison into anything that touches it, causing the familiar sting. A large number of plants have thorns, spines and prickles to defend them. These also help to reduce water loss.

---

er is always hot (except in ALPINE areas).

Tundra has the world's coldest recorded temperatures. It lies between the northern belt of coniferous trees and the Arctic circle. The main plants are lichens and mosses.

V **Venus fly trap** *(Dion- aea muscipula)* is a carnivorous plant found in California, USA. Its 2-lobed leaves are fringed with spikes. Also along the edges of the leaves are glands that produce nectar. Inside the spikes, the rest of the leaf is

covered by red digestive glands, and each lobe has three trigger hairs. A drop of rain that knocks one hair does not cause anything to happen but when an insect, attracted by the nectar, touches a trigger twice in succession, or touches two hairs, the trap is sprung. The lobes close rapidly, trapping the insect inside the inter-locked spikes, where it is digested.

W **Water hyacinth** *(Eich- hornia crassipes)* is a water plant that floats be-cause of pockets of air in its

Venus fly trap

leaf bases. It originated in the tropical regions of America, but is now causing serious congestion in many waterways.

Water lilies *(Nymphaea)* are plants with large floating leaves. They have attractive flowers and are much used in ornamental ponds.

Wolffia is the smallest known flowering plant, being only 0·5 to 0·7 mm across. It has no roots, stem or leaves, and consists merely of a tiny round body. Even so, *Wolffia* can cover the entire surface of a shel-tered pond.

X **Xerophyte** is a plant adapted to life in dry conditions.

Y **Yucca** is a genus of desert plants belonging to the lily family. They are found in Central and South America, where they are known as Joshua trees. Their flowers are pollinated by a particular species of moth. The female moth lays her egg inside the ovary of the flower and then deliber-ately pollinates it. The moth larva feeds on part of the fruit after it has developed.

All animals, including man, depend on plants for survival. The systematic, scientific cultivation of plants for food has allowed many species to thrive. However, man's progress threatens extinction for many species.

Plants and Man

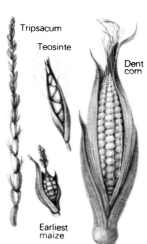

Left: The earliest maize had 48 kernels on each cob. It was cultivated by the Indians of tropical America. They cross-bred early maize with *Tripsacum dactyloides*, a wild grass relative. This produced teosinte (*Euchlaeana mexicana*), which they crossed again with early maize. The resulting larger cobs have since been selectively bred to produce modern maize, such as dent corn, which has 500-1000 kernels on each cob.

Tripsacum
Teosinte
Dent corn
Earliest maize

Early man was a wanderer. Small groups of people moved from place to place using what food they could find. Then, about 10,000 years ago, man discovered that the plant foods he needed would grow again in the same place from scattered seeds. He also found that he could help these plants to grow by planting the seeds in good soil. At the same time, if he saved the best seeds, the new plants would be stronger and more fruitful. So the descendants of early man settled down and became farmers. By selecting the best of their crops and animals for breeding, they began the process of domestication.

Underground food stores, such as roots, bulbs and tubers were probably the first to be used by man. They are available in almost every part of the world. Even today roots form the main diet for some primitive peoples, such as the few

Above: The 'wheat triangle' in Montana, USA, is an excellent wheat-growing area. The summers are hot and dry and the winters are cold and wet. Wheat is grown on about 30% of all the land used for grain crops.

Aborigines that still lead a wandering life in the Australian outback.

We also know that the pods and seeds of legumes (members of the pea family) were being eaten 10,000 years ago. The Indians of Peru ate KIDNEY BEANS 7,000 years ago, and broad beans have been part of the diet of Europeans for over 5,000 years.

The most important part of the early farmers' work was the cultivation of wild grasses. Over a period of thousands of years these grasses have been carefully bred into the modern cereals. Many of the original wild grasses are now extinct, but some still exist, and so we can trace the ways in which modern crops have been produced.

During most of this time the only method of breeding better plants was by selection. Farmers

Reference

A **Adder's tongue spearwort** (*Ranunculus ophioglossifolius*) is a rare British plant related to the buttercup. It is only known in one place — the Badgeworth Nature Reserve, Gloucestershire.

Almond (*Prunus dulcis*) is a small tree cultivated in southern Europe and Western Asia. It is grown for its edible seeds, contained in DRUPES (*see page 103*).

Apples are all derived from the wild crab apple (*Malus pumila*), a tree that grows in Europe, Asia and North America. After centuries of breeding there are now over 3,000 kinds of cultivated apples. The apple fruit is called a pome.

Artichokes are members of the daisy family. Globe artichokes (*Cyanara scolymus*) are closely related to thistles. They are grown for their flower heads, which are edible when immature. Jerusalem artichokes (*Helianthus tuberosus*) are related to the sunflower and are grown for their tubers.

Asparagus (*Asparagus officinalis*) is a member of the lily family grown for its edible young shoots.

Aster is a genus of peren-

Aubergine

nial garden plants. There are a number of hybrids, including the Michaelmas daisy (*A. novibelgii*).

Aubergine (*Solanum melongena*) is a perennial plant, also called the egg plant, grown for its fruit, which is a glossy, firm berry, either purple or white.

Aubretia is a genus of perennial garden plants with small purple flowers. They are small bushy plants frequently grown on walls or in rock gardens.

B **Banana** (*Musa*) is a genus of giant herbs grown in tropical regions. Cultivated varieties have edible, seedless fruits (berries).

Barley (*Hordeum*) is a genus of 6 species belonging to the grass family. The 3 cultivated species have 2, 4 or 6 rows of grain on the ears. Two-rowed barley is grown in Britain for making beer.

Beetroot (*Beta vulgaris*) is a purple root crop grown in many parts of the world.

Begonia is a large genus of garden and indoor plants. Some species are grown for their leaves, others are

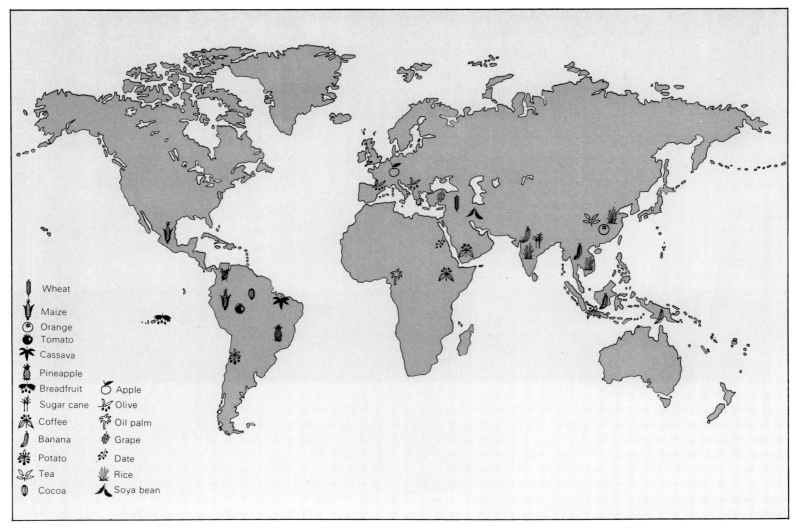

Wheat

Maize

Orange

Tomato

Cassava

Pineapple

Breadfruit   **Apple**

Sugar cane   **Olive**

Coffee   **Oil palm**

Banana   **Grape**

Potato   **Date**

Tea   **Rice**

Cocoa   **Soya bean**

Above: This map shows the origins of some of the world's main food crops. Some crops originated in Europe and Asia, but South America has also produced a large number.

and plant breeders merely selected the best strains and allowed them to grow. For example, MAIZE (sweetcorn) was produced in this way. The original plant had few seeds. By growing seeds from only those plants with the most seeds, farmers eventually produced the many-seeded maize plant grown today.

In the 1700s, scientists began to understand more about heredity – how characteristics are passed on from one generation to the next. Therefore, they could begin to use more scientific methods of producing food crops with better flavours, bigger yields and greater resistance to disease. One way of doing this is to cross-breed two different species. This is known as hybridization, and the plant that results from such a cross is called a hybrid. Many hybrids, particularly

animal hybrids, are infertile and therefore cannot be used to produce another generation, though a number of plant hybrids are fertile and are often much stronger than their parents. This strength is called hybrid vigour. Many of our modern cultivated plants, including ROSES, have been produced in this way. Hybrid orchids create spectacular displays of shape and colour.

Even more recently plant breeders have begun to use other ways of producing new plants. By treating seeds with chemicals or radiation (such as X-rays or gamma rays), they can cause changes in the chromosomes called mutations. These changes affect the nature of the plant. Most mutations are harmful, but some may result in better plants. Some varieties of wheat have been produced in this way.

grown for their flowers.
Bramble (*Rubus*) is a genus of rambling plants that includes blackberries, raspberries, loganberries and wineberries..
Broad bean (*Vicia faba*) is a member of the pea family that has large edible seeds.
Broccoli (a variety of *Brassica oleracea*) is a member of the CABBAGE family grown for its small flower heads. They are eaten together with the leaves.
Brussel sprout (a variety of the genus *Brassica oleracea*) is a member of the cabbage family that has large, dense

axillary buds. These are the sprouts that are eaten.

C **Cabbages** are varieties of *Brassica oleracea*.

Bramble fruit

They have very large terminal buds, which form the main bulk of the plant.
Carrot (*Daucus carota*) is a small biennial herb with a large edible tap root.
Cauliflower (a variety of *Brassica oleracea*) is a member of the cabbage family grown for its large white flower head, which is picked and eaten before it comes fully into bloom.
Cherry is a number of trees belonging to the genus *Prunus*. The red or purple fruits are edible drupes. The sweet cherry (*P. avium*) is the main source of fruit. Its

wood is used in veneers.
Coffee plant (*Coffea*) is a genus of small trees that are grown for their seeds. The fruits (berries) of the coffee

Wild cherry

plant are harvested, and the seeds, or beans, are used to make coffee.
Cotton plants (*Gossypium*) are 30 species of shrubs that grow in tropical and subtropical regions. They have large yellow, purple or white flowers. The fruits are capsules divided into compartments. Each compartment contains a seed surrounded by fine white fibres. These may be up to 5 cm in length.
Cretan date palm (*Phoenix theophrasti*) is a very rare palm tree found mainly in a single grove in Crete. It is threatened with extinction

Maize · Rice · Peanuts · Walnuts · Peas · Soya beans · Cherries · Gooseberries · Lemons · Peppers · Strawberries · Runner beans · Cauliflower · Broccoli · Artichoke · Asparagus · Sugar cane · Celeriac · Ginger · Potatoes · Spinach · Watercress · Radishes · Beetroot · Carrots · Turnips · Sugar beet

Plants for food

All the main parts of plants are represented in our food. From various plants we get seeds, fruits, flowers, stems, leaves and roots that can be eaten.

Edible seeds include those of the cereal crops, and the most important of these are RICE, WHEAT and maize. They are grown in many parts of the world and together they produce more than 750,000 million tonnes of grain each year. Other seeds that we eat include peas, soya beans and almonds. Coffee is made from the beans of the COFFEE PLANT.

A vast number of fruits are grown for food. As well as all the well-known fruits, such as peaches, oranges and blackberries, many of our vegetables are also fruits in the botanical sense. Examples of these include tomatoes, cucumbers and MARROWS.

Many flowers, too, are good to eat, although they are usually not allowed to bloom before they are picked. Cauliflowers, various kinds of broccoli and globe artichokes are all flowers.

The stems that we eat include not only above-ground stems but also tubers – modified underground stems. Asparagus, rhubarb and seakale are popular edible stems that grow above the ground. The stems of sugar can provide 65 per cent of the world's sugar supply. Cultivated tubers include potatoes, Jerusalem ARTICHOKES and yams.

Many plants are eaten for their leaves. The most obvious of these include cabbages, lettuces and leeks. ONIONS are bulbs and consist mostly of

Above: All the parts of a plant are represented in the foods we eat. The edible seeds, fruits, flowers, stems, leaves and roots of various plants are shown here.

---

through damage by tourists.
Cucumber, see MARROW FAMILY.
Cyclamen is a genus of 16 small garden and indoor plants related to the primrose. They come from the Mediterranean region and grow from corms.

D **Date palm** (*Phoenix dactylifera*) is a tall palm tree, about 25 metres high, that produces large numbers of single-seeded berries. It grows in dry sub-tropical regions.
Dracaena ombet is a palm-like tree that grows in Ethiopia and the Sudan near the Red Sea. It used to be common, but is now becoming rare. The local people use the trunks for firewood and the leaves for basket-weaving.

E **Ebony** (*Diospyros*) is a genus of trees that have black, heavy heartwood. They are found in tropical and sub-tropical regions. The best wood is obtained from the Ceylon ebony (*D. ebenum*).
Echium is a genus of plants found in the laurel forests of the Canary Islands. They are becoming rare, and this is partly due to the destruction of the forests by farmers.

F **Flax** (*Linum usitatissimum*) is a herbaceous

Cotton

plant cultivated for the fibres in its stem, in cool climates such as Europe, Russia and North America. The highest yield of fibre is obtained from plants harvested after the seed pods have ripened.
Freesia is a genus of greenhouse plants with trumpet-shaped flowers. They range widely in colour, and these plants can be grown from seeds or corms.

G **Gentian** (*Gentiana*) is a genus of perennial garden plants with blue or white flowers.
Gladiolus is a genus of monocotyledonous garden plants grown from corms. The flowers are borne on a long spike, and those at the bottom of the spike open first.
Gooseberry (*Ribes grossularia*) is a shrub grown for its green hairy fruits (berries). Currants also belong to the genus *Ribes*.
Grapefruit (*Citrus paradisi*) is a tree that grows about 10 metres tall. It has yellow-skinned fruits (berries) that grow in clusters.

H **Hemp** (*Cannabis sativa*) is the plant from which

Below: Honduras mahogany comes from one of the most important timber trees of tropical America. It is used in cabinet-making.

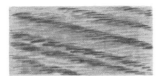

Below: Oak is a hard, heavy wood that wears well. It is used in the construction of homes and ships, and in making furniture.

Below: Cherry wood, grown in many parts of the Northern Hemisphere, is an attractive golden-brown wood used to make high-quality veneers.

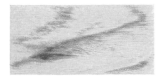

Below: Balsa wood, grown in tropical areas, is light and buoyant. It is used to make canoes, life-saving apparatus, and model aeroplanes.

Right: Opium poppies (*Papaver somniferum*) are grown in many parts of the world, especially in China, India and the Near East and Mediterranean area. About 2 weeks after the petals fall off, the poppy capsules are harvested for their opium, which is a white, milky juice.

Below: Ebony is a very hard wood. Ebony heartwood is especially valued for making inlays in cabinet work and for piano keys.

Below: Walnut is a tough wood with a particularly attractive grain. It is used to make fine furniture and interior panelling.

Below: Redwood wears well and is thus a useful wood for making garden furniture and siding for homes. It is also used for cabinet work.

Below: Rosewood is a fine reddish wood that takes a high polish. It is used in making furniture and veneers.

fleshy leaves. TEA is made from the leaves of a small tree.

Since man first discovered edible roots, many more plants of this kind have been produced. When settlers moved to new countries they took their crops with them. Root crops in particular were transported, because they are high yielding crops that are easily grown. As a result they are now found thousands of kilometres from where they first originated. Examples of widely-grown root crops are radishes, carrots, beetroots, and sugar beet.

Plants for materials and chemicals

For thousands of years, plants have provided man with many useful materials in addition to food. Shelter was the next most important need, and the earliest houses were built from mud and

grass. Even today, the common reed is still used to thatch the roofs of some houses in Britain. Later, when man had developed the appropriate tools, he used wood to build more solid houses. Today, few houses are built entirely of timber but wood is still used in construction and in making furniture.

There are two main types of timber. Softwoods are conifers, such as pine, spruce, and cedar. Hardwoods are broad-leaved trees, such as beech, oak, and WALNUT. The hardest woods of all are mahogany, TEAK, ROSEWOOD, and EBONY.

After a tree has been felled and cut into lengths, the wood is left to dry out, or season, for about a year. It is then suitable for use. As well as being used in building and furniture, wood is used to make telegraph poles and fencing, and it is also used as a fuel for domestic fires.

hemp fibres are obtained. It grows in sub-tropical regions. The drug marijuana is also obtained from this plant.
Hibiscadelphus is a rare Hawaiian plant whose flowers are adapted for pollination by the Hawaiian honey creepers. Due to the destruction of their habitat by man, both the plants and birds are in danger of extinction.
Hydrangea is a genus of 35 species of shrubs and climbing plants. *H. macrophylla* is a common shrub in British gardens. Its flowers may be pink or blue.

J **Jasmine** (*Jasminium*) is a genus of garden shrubs usually grown against walls.
Jute (*Corchorus*) is a genus of 2 species of annual plants

Gladioli

related to the lime, mostly found in Bangladesh. The stem fibres are coarser than those of FLAX and HEMP, but they used for making sacking and carpets.

K **Kidney bean** (*Phase-olus vulgaris*) is a member of the pea family. It is also called the French bean. Some varieties are grown for their edible fruits (pods). Others, such as haricot beans, are grown for their seeds only.

L **Lady's slipper orchid** (*Cypridedium calceolus*)

is the rarest British plant, known only in one place. It is also becoming rare in the rest of Europe.
Leek, see ONION.
Lemon (*Citrus limon*) is a small tree with yellow fruits (berries). Lemon fruits have less sugar than ORANGES, and thus are more acid to taste.
Lettuce (*Lactuca sativa*). A leafy salad vegetable related to the dandelion and daisy.
Love-in-a-mist (*Nigella damascena*) is an annual garden plant with finely-cut leaves and blue flowers.
Lupin (*Lupinus*) is a genus of about 300 species of her-

baceous plants. They are perennials and several are grown as garden plants.

M **Mahogany** (*Swietenia*) is a genus of trees

Hydrangea

Below: Teak is a hard wood when seasoned, and is popular for making veneers and furniture. It is also used in building ships.

Below: Ash, from Europe and North America, is a tough, springy timber used for making the handles of large tools, such as spades.

Below: Beech wood is valued for its hardness and smoothness. It makes good everyday furniture and handles for small tools.

Below: Pine is a popular light-coloured wood for making kitchen furniture and cupboards. It also has many other uses.

Below: The wood of the Douglas fir is often sold as 'Oregon pine'. The reddish-yellow timber is often used in construction.

Much of the softwood that is felled is made into paper. In this process the wood is first turned into pulp, which is done either chemically or by a machine. The pulp is refined by beating, which makes the fibres frayed and flexible. The grade of paper depends on this stage. Fibres beaten for a long time will produce the finest grades. Coarse papers, such as newspaper, do not need any further treatment. But the paper used in this book has been through several treatments to achieve this quality. Other kinds of treatment are used to produce grease-proof paper, cardboard, blotting-paper, tissues, and even paper clothes!

Wood is not the only product obtained from trees. Various species give us other materials essential to modern life. The bark of cork oaks is stripped off at intervals to give us cork – familiar to us as the stoppers of wine bottles. The RUBBER TREE is tapped for its milky white sap, called latex. This is treated by an industrial process to give us rubber. The saps of other trees supply us with natural resins and gums, such as resin from pine trees and canada balsam from the silver fir. Carnauba wax is obtained from the leaves of the carnauba palm. The waxy layer on the outside of the leaves is beaten off and is used in making polishes, crayons, cosmetics and carbon paper.

In addition to paper, other everyday materials are made from plant fibres. HEMP consists of fibres from the stem of the hemp plant. It is made into rope. The fibres of the sisal plant are also made into rope and string. Flax plants produce finer fibres that are used to make linen. Cotton, the finest of all the plant fibres, is obtained from the seed pods of the COTTON PLANT.

Some drugs and poisons are obtained from plants. The foxglove contains the poison digitalin. However, this can be used, in very small quantities, to treat some forms of heart disease as it makes the heart beat faster. Opium is a

Above: Rubber trees are tapped every few days by making a deep, sloping cut round the bark. The white, milky latex then seeps out until the wound heals. The latex is taken to a processing plant where it is turned into rubber by treating it with chemicals.

Below: The whitish wood from lime trees is soft but firm and it is used for carving. Lime trees are grown all over the Northern Hemisphere.

Below: Wood from spruces, which are conifers from the cooler regions of the Northern Hemisphere, is used for indoor woodwork, boxes, matches and paper.

Below: Yew is an increasingly popular, yellow-orange wood used in fine reproduction furniture, usually as a veneer.

Below: The wood of cedar trees is used for making cigar boxes, bottom planking in yachts and for lining clothing storage cupboards.

---

found in wet tropical forests. The 2 most important trees are the West Indian mahogany (*S. mahagoni*) and the Honduras mahogany (*S. macrophylla*).
Maize (*Zea mais*) is an annual grain crop principally grown in tropical and sub-tropical regions, but also grown in some temperate areas. There are a number of types, including dent corn, pop corn and flint corn.
Marrow family. Marrows, courgettes, squashes and pumpkins are all fruits (berries) of the genus *Cucurbita*. Many varieties are grown all over the world. Water melons (*Citrullus vulgaris*) are grown in many warm temperate, sub-tropical and tropical regions. Cucumbers

Oats

and other melons are fruits of the genus *Cucumis*.

O **Oats** (*Avena sativa*) are grown as a grain crop in temperate regions. They can easily be distinguished from other grain crops by their spreading flower head with hanging spikelets — quite unlike the tight flower heads of wheat, rye and barley.
Onion (*Allium cepa*) is a vegetable widely grown from a bulb, particularly in Egypt, Europe and north America. There are a number of varieties grown for cooking, pickling or eating raw in salads. Leeks, garlic and chives also belong to the genus *Allium*.
Orange. Several varieties of tree bear orange fruits (ber-

Hemp

ries), including sweet oranges (*Citrus sinensis*) and Seville oranges (*C. aurantium*).

P **Pea** (*Pisum sativum*) is an annual plant grown for the seeds contained in its pods.
Peach (*Prunus persica*) is a small willowy deciduous tree that bears velvety-skinned fruits (drupes).
Pears include a variety of trees all descended from the common pear (*Pyrus communis*), a tree that grows wild all over Europe. The fruits (pomes) are formed in

Swiss cheese plant

Orange tree

Ivy

Rubber plant

Poinsettia

Christmas cactus

Alpine violet

Tradescantia

Begonia Rex

Busy Lizzie

Lady's pocket book

Mexican sunball

dangerous drug obtained from the unripe seed capsules of the opium poppy. It can be used to make codeine, which is a mild pain-killer, or morphine and pethidine, which are powerful pain-killers. Cocaine is a drug that acts like an anaesthetic. It is obtained from the cocoa plant. Quinine is extracted from the bark of the cinchona tree. This drug was once used to treat malaria, but it has now been replaced by other drugs. Caffeine is a drug, found in tea and coffee, that acts as a stimulant. The leaves of the tobacco plant contain nicotine, which is another drug that stimulates the nervous system.

The deadly poisons found in plants include atropine from deadly nightshade, and coniine from hemlock. Curare is a mixture of deadly drugs. In the forests of South America it is used to tip blow-gun darts and arrows for hunting.

Above: There are many varied indoor plants. Some are grown for their attractive foliage, others for their showy flowers or fruits, and some just for their unusual appearance.

Plants for pleasure

As civilizations grew, people had time to relax and enjoy the things around them. Nature has her own beauty, but for thousands of years man has delighted in using flowers to create beautiful gardens.

The Bible tells us of the Hanging Gardens of Babylon, which the Greeks described as one of the seven wonders of the world. Since then gardens have reflected the tastes of society. The ruins of Pompeii show us the small enclosed gardens of the Romans. Much later, in the 1600s, the gardens of the great European houses and palaces were extremely formal, and were laid out with great care. In the 1700s there was a reaction to this formal style, and men such as Capability Brown changed gardens into 'natural' landscapes.

the same way as APPLES.
Plums include a number of fruits (drupes) of trees belonging to the genus *Prunus*. Also in this group are damsons, gages, sloes, and bullace.
Potato (*Solanum tuberosum*) is a plant of the nightshade family grown for its tubers. It is one of the most important food plants of the world, and there are several varieties.

R **Radish** (*Raphanus sativus*) is a widely grown root crop used in salads.
Rhubarb (*Rheum rhapon-*

ticum) is a perennial plant grown for its leaf stalks. The leaves are not eaten as they contain oxalic acid poison.
Rice (*Oryza sativa*) is the

Rice harvest in Guyana

chief cereal crop of Asia, developed from a marsh plant and adapted to growing in flooded areas. Rice seed is sown in nurseries and the seedlings later transplanted into flooded fields. A few weeks before harvesting the fields are drained to allow the ears to dry off and the grain to ripen.

Roses (*Rosa*) are probably the most popular of all garden plants. There are several types, including bush, shrub, standard, rambling, climbing and miniature roses. Breeders are constantly trying to pro-

duce new varieties with ever more beautiful blooms.
Rosewood (*Dalbergia*) is a genus of evergreen trees that grow about 10 metres

Sugar cane harvest

tall. They all have attractive wood, and the most important species are the Brazilian rosewood (*D. nigra*), the Honduras rosewood (*D. stevensonii*), the Indian sissoo (*D. sissoo*) and the East Indian rosewood (*D. latifolia*).
Rubber tree (*Hevea braziliensis*) is a tree that is more accurately called the Pará rubber to distinguish it from the rubber tree (*Ficus elastica*), an Asian tree commonly grown as a houseplant. The Pará rubber comes from Brazil, but most rubber plantations are now

Bottom right: Roses are one of the most popular garden plants. They have beautiful flowers, which are often sweet-smelling. Rose breeders spend much time in establishing new varieties, and roses are grown in several forms.

Right: Some gardens are very formal. This style of garden is popular in Japan, where such settings are used for quiet relaxation and contemplation.

In the past gardening was largely a rich man's pastime, but today nearly everybody has a garden of some kind. It may be a large country garden or a small patch at the back of a town house. The style may be formal, informal or a mixture of both, and gardens now reflect the tastes of their individual owners. The wide range of plants and garden materials now available make it possible to create any style of garden desired.

All the garden flowers, shrubs and trees are descended from wild plants. Careful selection and breeding, including hybridization, has being going on for over 3,000 years, since the Egyptian and Assyrian civilizations. The skills required to tend and cultivate plants were passed on to the Greeks and Romans. During the Dark Ages, monks used these skills to cultivate garden plants in the privacy of their cloisters. At the same time the Chinese and Japanese gardeners were developing their own gardening techniques, and plants were also being cultivated in Central and South America. Modern gardens contain plants from all over the world, and we owe this to the great European explorers of the 1400s to 1800s who brought back unknown seeds, bulbs and living plants from their voyages. Some of these did not survive in their new surroundings, but many of them have become successful and popular garden plants. For example, chrysanthemums originated in Japan, pelargoniums and gladioli came from South America, and hyacinths were first bred in Asia. Subsequent breeding and selection has produced all the different varieties of these plants that we know today.

Plants in danger

The number of animals that are threatened with extinction is increasing, and there is much worldwide concern about this. However, because

found in Malaysia.

Runner bean (*Phaseolus coccineus*) is a climbing plant of the pea family, also known as the scarlet runner,

Young teak trees

that is grown for its fruits (pods).

Rye (*Secale cereale*) is an important grain crop in the colder regions of Russia and Europe. It is used for making bread, gin and beer, and the young shoots are sometimes fed to animals.

S Seakale (*Crambe maritima*) is a seaside plant related to the wallflower. If it is grown in the dark, its leaf stalks remain white and can be eaten as a vegetable.

Sisal (*Agave sisilana*) is a succulent plant related to the American century plant. It contains tough fibres in its leaves that are extracted and used to make rope and string.

Soya (*Glycine max*) is an annual plant of the pea family grown for its seeds. Soya beans are used to make an edible oil, or they can be eaten as a vegetable. Soya beans contain a high proportion of protein and are therefore a useful source of food.

Strawberry (*Fragaria*) is a number of cultivated and wild plants, mostly grown in Europe and America. The juicy, red edible structures are in fact false fruits, because they are formed from swollen receptacles. The true fruits of the strawberry are the achenes on the out-

Tea plant

side of the edible part.

Sugar beet (*Beta vulgaris*, sub-species *cicla*) is a root crop closely related to the BEETROOT. It is grown in Europe, North America and Russia, and is an important source of sugar in countries too cold to grow sugar cane.

Sugar cane (*Saccharum officinarum*) is a grass whose stems, or canes, contain a large amount of sugar. It is grown in tropical regions, where it gives the best sugar yield, and in sub-tropical areas.

Sweet pea (*Lathyrus odoratus*) is an annual

areas increases the chance of soil erosion, which leaves the land bare and useless.

Plants are also being threatened by the activities of man in many other areas of the world. Overgrazing dry areas has led to the formation of deserts, such as the belt from the Sahel in Africa to south-western Asia. Goats in particular are responsible for such effects, as they eat anything green, including young tree shoots.

Many of the world's islands have also suffered. On Hawaii there are 800 endangered species of plants, and 273 are listed as extinct. St Helena in the South Atlantic has been affected by both grazing animals and plants introduced from outside. Goats, first introduced in 1513, devastated the forests within 100 years. In 1805 a botanical expedition discovered 31 species of native plants. Eleven of those are now extinct. The goats are now under control, but two introduced plants, New Zealand flax and gorse, are causing more problems. They are spreading over the island, swamping the natural flora.

it is less easy to feel sympathy for a plant, we hear much less about the plants that are also becoming rare. In fact about 25,000 species of plants (ten per cent of the world's flora) are nearly extinct.

We should be concerned about this because animals and man depend on plants for survival. If areas of plant life are destroyed, then the animals that live in those areas will die out. At the same time many plants that might be useful to man are also being destroyed. This is particularly true of the tropical rain forests. When areas of these forests are cut for timber or cleared for agriculture, the plant and animal populations are reduced in numbers. Some localized plants are wiped out altogether. In the Phillipines, 172,000 hectares of forest are being removed every year. In the Amazon basin land is being cleared for development. In this relatively unexplored area we are probably losing many species of plants before they have even been discovered. Some of those plants could, perhaps, have provided us with new sources of food, new drugs or new materials, such as oils.

This could be described as one of the penalties of progress but many people believe that the land is actually unsuitable for grazing or growing crops. In addition, removing forests from hilly

Above: The lady's slipper orchid (*Cypripedium calceolus*) is the rarest plant in Britain, and is becoming rare all over the Northern Hemisphere.

Right: The silver sword (*Argyroxiphium macrocephalum*) is only found on a crater of the volcanic Maui Island – one of the Hawaiian Islands.

---

climbing plant that produces a great show of sweet smelling flowers.
Syagrus sancona is an endangered species of palm tree that grows in the rain forests of Columbia. Areas of this rain forest are now being cleared for grazing.

T **Teak** (*Tectona grandis*) is a large tree that grows in India and Burma. Its heartwood is golden yellow after the tree has been cut, but becomes brown and mottled when seasoned.
Tea plant (*Camellia sinensis*) is a small tree grown in

tropical and sub-tropical areas of the world. Tea is made from its leaves, and the finest tea is made by using only young shoots.

Walnuts

Tobacco plant (*Nicotiana tabacum*) is an annual plant belonging to the nightshade family. It is grown in many parts of the world, and its leaves are harvested and cured to make tobacco.
Tomato (*Lycopersicon esculentum*) is an annual plant belonging to the nightshade family grown for its red or orange fruits (berries).

W **Wahlenbergia linifolia** is a small shrub with white bell-shaped flowers that grows on the island of St Helena. Because it is small it is in danger of being

swamped by the plants that have been introduced to the island by man.
Wallflower (*Cheiranthus*) is a genus of popular biennial

Yams

garden plants.
Walnut (*Juglans*) is a genus of 17 species of deciduous trees. The best nuts (seeds) are obtained from the fruits (drupes) of the English walnut (*J. regia*).
Wheat (*Triticum*). Several species of this grain crop are grown in Europe, Russia, Asia, America, and Australia. It is used for making flour for bread, biscuits and pasta.

Y **Yam** (*Dioscorea*) is a genus of climbing plants grown for their edible tubers in America, Asia and Africa.

Index

Acknowledgements

Contributing artists
Terry Callcut, John Goslar, Tim Hayward, Ron Haywood,
Kate Lloyd-Jones, Elaine Keenan, Abdul Aziz Khan

The Publishers also wish to thank the following:
A–Z Botanical Collection 68B, 82B, 97BL BR, 99B, 102L, 104BL BR, 98L, 100B, 105L, 107BL BR,
108BR, 110, 122R, 124BL, 128BL
Heather Angel 30, 32BL, 33B, 35B, 36B, 37B, 38B T, 40TL, 46TL, 48, 54TR, 55, 56, 73TR, 77, 80BL BR,
83TR, 93BL, 95, 100, 103TL, 104, 106R, 107, 108BR, 114, 119T, 120L
Aquila Photographics 13B, 31B, 51BL, 78B, 86B
Ardea 115C, S. Roberts 21C, K. Fink 23TL, B. Stonehouse 25T, D. Burgess 28TR, C. Weaver 61R, W.
Weisser 68T, Clem Haagner 83CL, I. Beames 103TR, Dr. Pat Morris 125T
Edward Ashpole 3
Barnaby's Picture Library 61BL
Biophoto Associates 67, 73C, 74, 75T, 76, 87CR, 91, 108TL TR, 109TL TR, 119T
Anne Bolt 121B, 126L R, 127BL, 128BR
Douglas Botting 62TL
Bruce Coleman Ltd. 29T, 61TL, 121T, P. Arnold 26T, Jen and Des Bartlett 22T, 90, S. C. Bisserot 94,
Jane Burton 8L, 14TL, 15, 37TR, 43T, 46B, 119C, Gerald Cubitt 97T, Stephen Dalton 18, 24T,
Francisco Erize 12BC, Gordon Langsbury 58T, Norman R. Lightfoot 63BL, John Markham 4T,
Norman Myers 62TR, Oxford Scientific Films 83CR, J. M. Pearson 16T, G. D. Plage 13, Allan Power
38CR, Hans Reinhard 32CL, 128TL, Norman Tomalin 34TR, John Wallis 14BR, 31C, Bill Wood 36TR
Robert Estall 81
Fisons 85
Brian Hawkes 36TL, 54C, 70TL, 71R, 128C
Eric Hosking 17L, 20BL, 21BL BR, 22B, 23TR, 54B, 55BL
Sarah King 41L
Mark Lambert 111C, 127C
Pat Morris 8R, 16BL, 21T, 40CR, 41R, 54TL
Natural History Photographic Agency 10TL, 12T C, 16C, 27T, 28BL, 31T, 32CR, 33T, 34CL, 40TR, 47R,
60T, 63TR, 80TR, 89T, 112T C, 117C
Natural Science Photos 19, 28TL, 37TL, 86T, 87CL, 89C, 109TR, 118L, 124T, 127T
Popperfoto 120B
Radio Times Hulton Picture Library 4B, 6, 7, 70BL
G. R. Roberts 45, 51BR, 71L, 96, 97C, 99T, 113T, 115T BR, 120TR
Harry Smith Horticultural Photographic Collection 78T
Spectrum Colour Library 10TR, 20, 44T, 50, 51T, 70R, 87TR
Peter Stiles 117T
Thames Water Authority 82T
John Topham Picture Library 5, 9, 10BL BR, 11L R, 12B, 15B, 16BR, 17R, 19B, 20BR, 23BL, 24B, 25B,
26B, 27B, 28BR, 29B, 30B, 32B, 34B, 40BL BR, 42, 43BL BR, 44B, 49, 52, 53, 55BR, 56B, 57, 58BL BR,
59, 60B, 62BL BR, 63BR, 64, 72L R, 73B, 75B, 79, 87BL BR, 88, 89BL BR, 92, 93BR, 95B, 98R, 101L R,
102R, 103BL, 105R, 106C B, 108BL, 109B, 111B, 112B, 113B, 114B, 115BL, 116, 117BL BR, 118B, 119B,
122L, 123 124BR, 125BL BR, 127BR
Douglas P. Wilson 47L
World Wide Butterflies 44C
Zefa Picture Library 35T